Tissue Repair and Reconstruction

Series Editors

Andy H. Choi, Carlingford, Australia

Besim Ben-Nissan, Sydney, Australia

SpringerBriefs in Tissue Repair and Reconstruction provides a unique perspective and in-depth insights into the latest advances and innovations contributing to improved and better treatments for patients with damaged soft and hard tissues as a result of diseases, trauma, and implantations. The book series consists of volumes that offer biomedical researchers better insights into the advancements of biomaterials science and their translation from the laboratory to a clinical setting. Similarly, the series provides information to surgeons and medical practitioners on novel ideas in biomedical science and engineering on top of disseminating new ideas and know-hows in diagnostics and treatment options for patients from head to toe.

The series will cover a number of key topics:

Fundamental Concepts and Surface Modifications: The topic will provide detailed information on the discovery and advancements of biomaterials surface modification approaches and their use within the human body in a safe manner and without provoking any negative tissue response.

Computational Simulations and Biomechanics: Anatomically accurate computational models are being in all fields of medicine particularly in orthopedics and dentistry to reveal the biomechanical functions and behaviors of bones and joints when damaged, diseased, and in the health state. They also contribute to our understanding during the design and applications of implants and prosthetics subjected to functional loadings and movements.

Surgical Advances and Treatment Options: Discusses how surgical techniques are revolutionized by our deeper understanding into biomaterials science and tissue engineering. The section also focuses on the latest innovations and surgical advancements currently being used to treat patients with damaged tissues.

Post-Operative Treatment and Rehabilitation Engineering: Expands the independence and functionality of the patient after surgery while at the same time reducing the chance of complications such as wound infections and dislocations. Advances in technologies are creating new opportunities in how physiotherapy rehabilitations are delivered.

Necdet Sağlam · Feza Korkusuz · Mesut Şam
Editors

Nano-Biomaterials in Tissue Repair and Regeneration

Surgical Aspect for Hard Tissues

Springer

Editors
Necdet Sağlam
Nanotechnology and Nanomedicine Division
Hacettepe University
Ankara, Türkiye

Feza Korkusuz
Department of Sports Medicine
Hacettepe University Medical Faculty
Ankara, Türkiye

Mesut Şam
Department of Biology
Aksaray University
Aksaray, Türkiye

ISSN 2731-9180 ISSN 2731-9199 (electronic)
Tissue Repair and Reconstruction
ISBN 978-981-96-3807-9 ISBN 978-981-96-3808-6 (eBook)
https://doi.org/10.1007/978-981-96-3808-6

This Springer imprint is published by the registered company Springer Nature Singapore Pte Ltd.
The registered company address is: 152 Beach Road, #21-01/04 Gateway East, Singapore 189721, Singapore

Preface

Clinical regeneration of bone and joint cartilage tissues is currently moving to a new dimension. Regenerative medicinal approaches are advancing by innovative research and technological breakthroughs in nanotechnology and new biomaterials. Advanced technology in omic studies as well as artificial intelligence has started to dominate our research journeys. This book aims to provide a comprehensive overview of the latest developments in the research and application of nano-biomaterials for the repair and regeneration of hard tissues.

Regeneration of bones that are common after gunshot wounds, high energy road traffic and workplace accidents emerge with complex and open fractures. The incidence of infection has to be prevented and/or treated. Vascularization should be re-established. Fractures in osteoporosis is another challenge in an aging population. Traditional methods of treatment, while effective to some extent, often fall short in terms of long-term outcomes and functionality. This is where nano-biomaterials come into play, offering novel solutions that enhance the efficacy and success of tissue repair and regeneration. Nano-biomaterials, due to their unique properties at the nanoscale, can mimic the natural extracellular matrix, promote cellular activities and provide mechanical strength to biological structures. Their ability to integrate with biological systems at the molecular level makes them ideal candidates for developing advanced therapeutic strategies.

We explore nanoparticles, nanostructures and smart piezoelectric materials of various types and their roles in hard tissue repair and regeneration. We furthermore give insights into omic research and artificial intelligence as they interact with biological tissues and revolutionize clinical practices.

Our goal is to provide recent researcher outcomes to clinicians and students with a thorough understanding of future directions of nano-biomaterials in hard tissue repair and regeneration. We hope that this book will inspire further research and innovation, ultimately leading to improved patient outcomes and the advancement of medical science.

Ankara, Türkiye Necdet Sağlam
Ankara, Türkiye Feza Korkusuz
Aksaray, Türkiye Mesut Şam

Contents

Craniofacial Bone Defects, Cranial Bone Tissue Engineering, and Reconstruction

Alaz Çırak and İbrahim Vargel

Abstract Craniofacial bone defects occur due to various reasons such as congenital developmental disorders (craniosynostosis, cleft lip-palate etc.), traumas and oncologic resections. The main goals of craniofacial reconstruction are to provide the patient with a proper craniofacial form, to regain function if there is loss of it, to eliminate aesthetic concerns, and to ensure the continuity of craniofacial development in a normal way for the pediatric patients. Because there is an underlying bone defect, the use of bone grafts is often indispensable in craniofacial reconstruction. As the gold standard, autologous bone grafting is the first choice under ideal conditions. However, due to limited autologous graft sources, additional surgical intervention requirements, and resulting donor site morbidities; craniofacial reconstruction has necessitated alternative solutions that combine tissue engineering and regenerative medicine. Growth factors such as bone morphogenic protein (BMP), which effect bone and cartilage formation and regulate bone healing physiology, have attracted attention in recent years. As a result of recent advances in tissue engineering and regenerative medicine; cells, tissue scaffolds, and growth factors are used together due to the importance of 3D tissue healing in craniofacial defects, and the complex nature of the tissue repair mechanism.

Keywords Bone replacement material · Bone substitute · Craniofacial abnormalities · Maxillo-mandibular reconstruction · Prostheses and implants · Reconstructive surgical procedure · Regenerative medicine · Stem cells · Tissue engineering · Tissue grafts

---

A. Çırak
Department of Plastic Reconstructive and Aesthetic Surgery, Faculty of Medicine, Hacettepe University, Ankara, Turkey

İ. Vargel (✉)
Bioengineering Division, Hacettepe University Institute of Science, Ankara, Turkey
e-mail: ivargel@gmail.com

N. Sağlam et al. (eds.), *Nano-Biomaterials in Tissue Repair and Regeneration*, Tissue Repair and Reconstruction, https://doi.org/10.1007/978-981-96-3808-6_1

1 Introduction to Craniofacial Bone Defects

Craniofacial skeleton consists of cranial bones and facial bones. Craniofacial bone defects are the areas of bone loss at any part of craniofacial skeleton [1]. Defects can occur due to various reasons, such as trauma, infection, neoplasia, or congenital malformations [2] (Fig. 1).

Craniofacial bone defects often cause disturbance of basic functions such as seeing, breathing and feeding. Fractures and defects that concern the orbit threaten vision [3], maxilla-mandibular fractures or defects that compromise airway [4] or oral-oropharyngeal cavity disturbs breathing and nutrition. Craniofacial defects can cause significant disturbance of the patients' appearance [5] and these defects should be correctly reconstructed. Craniofacial reconstruction is one of the main topics of plastic and reconstructive surgery [6]. Defects in craniofacial skeleton can be divided as cranial and facial defects.

Cranial bone reconstruction or cranioplasty is one of the most ancient surgical techniques with first reports in are from 3000 BC, being reconstruction of trephination defects with gold and silver [6]. Cranial bone reconstruction indications accepted in the present day are treatment of aesthetic contour deformities, protection of underlying brain tissue, treatment of the "trephine syndrome" which is caused by pressure and circulation changes of the central nervous system [7–9]. Indications of facial bone reconstruction are similar to indications of cranial bone reconstruction, main indications are restoration of function and treatment of aesthetic contour deformities [10].

Reconstruction of calvarial defects can be complicated with infections. In order to reduce this complication, the timing of cranial reconstruction has been of debate [11, 12]. Some studies argue that in order to reduce infectious complications reconstruction should be performed 90 days after formation of the defect [13]. On the other hand, there are also studies that show no difference between complication rates of instant or delayed cranial reconstruction [14]. Early cranioplasty may also be related to better neurological outcomes [15].

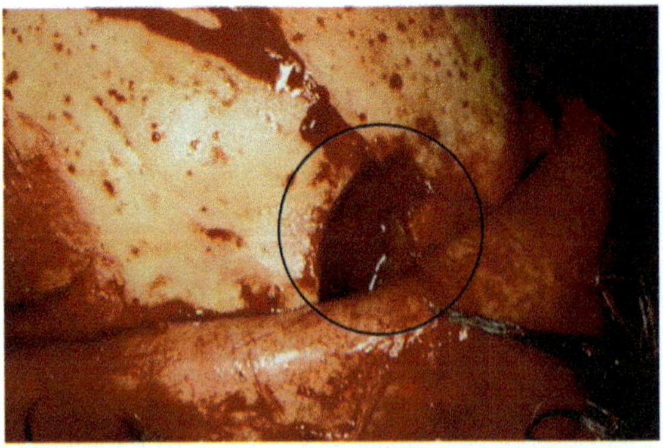

Fig. 1 Cranial bone defect. A traumatic frontal bone defect seen in an intraoperative photograph. Defect is marked with a circle. Upper side of the picture is superior/cranial and lower side of the picture is inferior/caudal of the patient

Craniofacial defects can be reconstructed with autologous or alloplastic materials [16]. In autologous reconstruction, the defect is reconstructed with tissues of the patient such as bone grafts or flaps. Reconstruction with autologous bone graft has been considered the gold standard method when treating cranial bone defects especially in pediatric patients [17]. However, in certain cases, autologous reserves can be insufficient, in order to overcome reserve problems and prevent possible donor site complications alloplastic materials can be used in reconstruction of craniofacial defects [10].

2 Autologous Reconstruction of Craniofacial Bone Defects

Following certain procedures such as decompressive craniectomy, the underlying cranial bone can be stored in freezers or can be implanted in a different place of the body [18–20]. If such a bone flap is available, it is considered the primary method in reconstruction of cranial defects [21].

Despite the advances in biomaterials, the gold standard method in reconstruction of craniofacial defects remains the autologous bone grafts [22, 23]. Autologous bone grafts have lower infection rates, fewer late-term complications, and are safer among pediatric patients since they allow cranial growth [22].

Main donor sites are cranium, iliac bone, and ribs. Use of cranial bones as the donor site in reconstruction of cranial defects is favorable as cranial bones are easily accessible from the original incision [1] (Fig. 2). The most commonly preferred donor site is the right parietal bone [10]. Donor cranial bone is split from the diploic space, inner table is left in the donor site and outer table can be used as the bone graft and placed to the primary defect and fixed to the neighboring bones [21]. Small dust like bone fragments can be harvested from cranial bones as a bone graft (Fig. 3) [24].

Iliac bone is the preferred donor site for bone grafting in orbital floor reconstruction and in cases where high amounts of cancellous bone is necessary [1, 21].

Bone defects in craniofacial skeleton can be reconstructed with bone grafts. Bone grafts are used in reconstruction of cranial bone defects, nasal bone defects, defects of the nasoethmoid or orbitozygomatic areas, correction of posttraumatic enophthalmos, hypoplastic irradiated orbit, maxilla, and mandibular reconstruction [10, 21].

Following the administration of bone grafts, each surface of the graft must be covered with well vascularized recipient tissues. Furthermore, if the defect is greater that 4–6 cm, vascularized free bone flaps can be considered such as iliac or fibular bone flap [25].

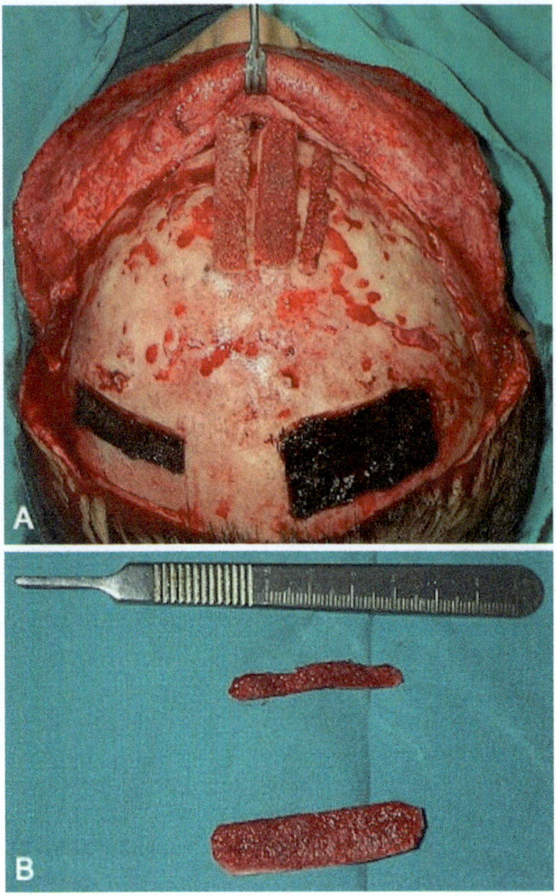

Fig. 2 Cranial bone graft. Harvested cranial bone graft during a nasal encephalocele correction operation can be seen in the photograph. **a** Cranial bone grafts are harvested from outer table of parietal bones. Hemostatic materials can be seen on donor sites of the bone grafts. **b** Harvested bone grafts can be seen

3 Allogeneic Reconstruction of Craniofacial Bone Defects and Bone Xenografts

Although autologous bone grafting is the gold standard in reconstruction of craniofacial bone defects, donor site morbidities and limited resources are main downsides. The main alternatives are allogeneic bone grafts or demineralized bone matrices (DBM) [26]. Allogeneic bone grafts are essentially bone grafts from a different donor and DBM's are bone processed grafts from another individual [27]. Allogeneic bone grafts are osteoconductive, osteoinductive and have great osteogenic potentials similar to autogenic grafts however, deproteinization which is necessary to decrease immune reactions, decrease their osteogenic potentials [28–30].

Bone allografts can be divided as fresh vs frozen or cortical vs cancellous. Bone allografts can be harvested from various bones of a deceased donor, such as cranial bones (Fig. 4), iliac bones, or long bones like tibia and fibula. In order to

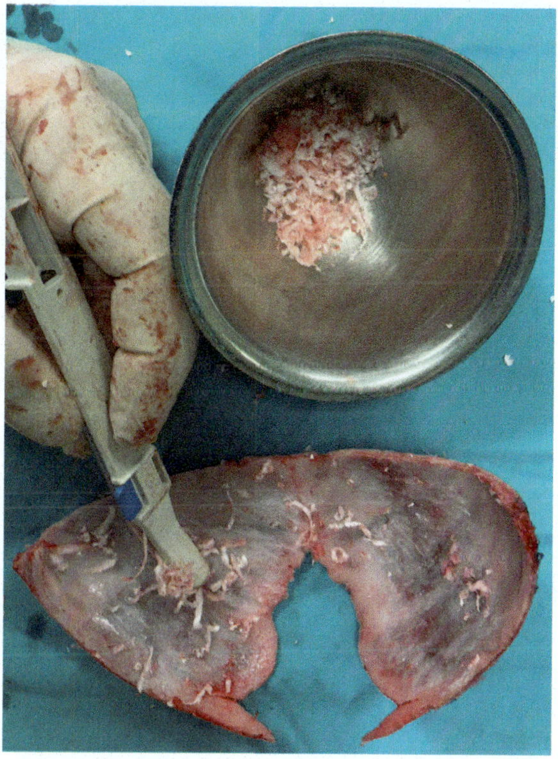

Fig. 3 Autologous bone dust. Autologous bone dust harvest from inner surface of an harvested autologous bone graft can be observed. Separated bone dust fragments that are placed in a bowl can be seen in the upper side of the picture

decrease the possibility of host reactions, allografts can undergo specific preparations such as debridement, disinfection, irradiation, cryopreservation, or they can be used as unprocessed [26]. Two possible downsides are as follows: Allografts can be unsafe regarding transmission of infectious diseases similarly to autographs, there are limited sources of allografts [31].

When allografts are processed in a specific way that was defined by Urist, DBMs are formed [32]. With this process, inorganic mineral components are removed and remaining organic components of the bone including collagen matrix, growth factors are left in the graft [26, 33]. DBMs can be designed in various types, the most commonly used type is a moldable bone paste type and this moldable paste can be coupled with biocompatible viscous carriers such as polymer carriers (chitosan, hyaluronic acid, sodium alginate, corboxtmethylcellulose, type 1 collagen, poloxamer 407) or low molecular weight carriers (glycerol, calcium sulfate, bioactive glass) [34, 35]. DBM can be used as a scaffold for bone tissue engineering [36] or in combination with autografts [37].

Similar to bone autografts, allografts also have limited availability; therefore, xenografts and xenograft-derived scaffolds have been studied [38, 39]. Bovine [40–42] and porcine [40, 43] bone transplantation have been studied. Xenografts carry certain transmission and rejection risks similar to allografts and a xenograft can be

Fig. 4 Cranial bone allograft. **a** Cranial bone allograft can be seen. Allograft is being separated with a bone saw. **b** Separated inner and outer tables of the allograft can be observed

used if processing chemicals are completely removed, infectious transmission risk is minimized and immune reactions and rejections are minimized [36].

4 Reconstruction of Craniofacial Bone Defects with Implants

Implants overcome the donor site morbidities of autologous reconstruction and provide limitless amount of graft material [1, 44]. Alloplastic materials are favorable in cases where resorption of the grafted tissue cannot be tolerated. The qualities of an ideal bone implant are [44–46]:

- Biologically inert, causes minimal foreign body reaction
- Elastic
- Easy to produce
- Good tissue incorporation
- Allows collagen ingrowth
- Promotes permanent tissue repair
- Can tolerate infected environment
- Stable, durable
- Has minimal complications

Metals, polymers and ceramics have been used as alloplastic implants for bone reconstruction.

4.1 Metals

Metals have been used as implant materials and metal alloys have been developed in order to reduce their corrosion and increase inertness [47, 48]. Main metals and metal alloys are stainless steel, cobalt-chrome, titanium, gold, and platinum.

4.1.1 Stainless Steel

Stainless steel is an alloy made of iron-chromium-nickel. It is developed in order to overcome corrosion. Its tensile strength is high and it is deformable [49]. It is used in arch-bars [50]. One drawback is that the metal ions released to the environment and it causes inflammatory reactions in the host [44].

4.1.2 Cobalt-Chromium

Cobalt is mixed with chromium and molybdenum to form vitallium alloy [48]. Vitallium is used in craniofacial reconstruction as miniplates and screws. The major drawback is that it causes great artefacts in computerized tomography [51]. Therefore, it was replaced by, titanium.

4.1.3 Titanium

Titanium alloys are more durable, stronger, higher anti-corrosion properties, more biologically inert and have lesser stress shield effect on bones [48]. It has high osseointegration potential, it chemically interacts with neighboring bone. Titanium alloys are used in craniofacial reconstruction as plates, screws and as meshes [52]. Plates made of titanium can be used in correction of orbit floor fractures (Fig. 5). Titanium implants are widely used in craniofacial reconstruction especially in low income and low resource settings in the absence of customized implants and synthetic biomaterials [53].

4.1.4 Gold

Gold is biologically and chemically inert [54]. However, it is physically weak therefore it can be used in alloys. Gold is used in eyelid weights of patients with lagophthalmos [55].

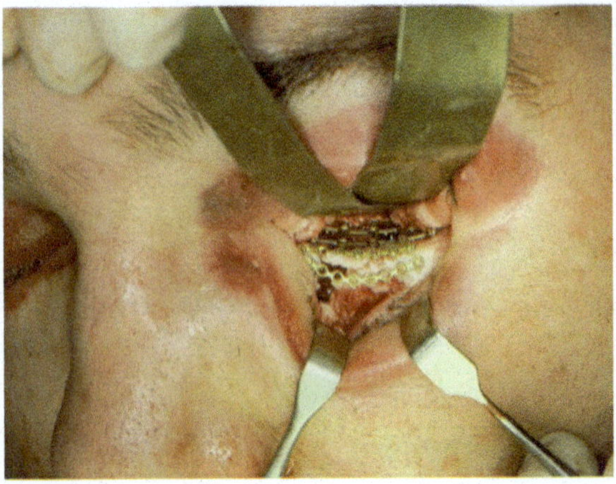

Fig. 5 Titanium mesh in orbit floor fracture repair. A titanium microplate placed at the orbit floor can be observed in the picture. Two titanium plates can be seen placed at the infraorbital rim

4.1.5 Platinum

Platinum is biologically inert and is used as eyelid weights for patients who cannot tolerate gold weights [56]. The density of platinum is higher than gold therefore, smaller weights can be designed [44].

4.2 Polymers

Polymers are made of repeating monomers and have cross-links in their structures. Physical and chemical properties of polymers change as number of monomers and cross-links change.

4.2.1 Silicone

Silicone is a polymer made of silica (Si) and oxygen (O) molecules, and repeating molecule is called Siloxane. Silicone is used for medical purposes in poly-dimethylsiloxane (PDMS) form which is chemically and biologically inert [57]. Silicone implants are used in craniofacial surgery for only implants and for orbital floor reconstruction [58].

4.2.2 Polytetrafluoroethylene

Polytetrafluoroethylene (PTFE) implants are polymers made of carbon and fluorine. It is porous form so that host tissue can integrate into the implant to some extent.

PTFE implants are used in craniofacial reconstruction for contour augmentation, PTFE is also used in meshes [59].

4.2.3 Polyester

Polyester is a polymer containing an ester group. It has favorable tissue ingrowth properties. It is used in meshes [60].

4.2.4 Polypropylene

Polypropylene (or polyprolene) is a polymer made of carbon hydrogen and methyl groups. It is inert, strong and it causes low foreign body reactions. It is used in meshes and in surgical sutures [61].

4.2.5 Polyethylene

Polyethylene is a polymer made of carbon and hydrogen. Medpore which is commonly used as a craniofacial implant is a polyethylene that it has a porous form and a high density [62]. Its porous form enables tissue integration, tissue ingrowth. Due to high tissue ingrowth removal of the implant is challenging. One disadvantage is that the implant is not visible on computerized tomography [63]. High density porous polyethylene is clinically widely used and widely studied in alloplastic cranioplasty and in recent studies their use was associated with low complication rates with high patient satisfaction [64].

4.2.6 Polyether Ether Ketone (PEEK)

Polyether ether ketone (PEEK) is a polyaromatic biologically inert polymer that has a semi-crystalline and thermoplastic structure. PEEK implants have demonstrated high strength and elasticity and low thermal conductivity and infection rates in previous studies [65]. PEEK implants have been used in craniofacial reconstruction [66], often with CAD/CAM technology as patient and defect specific implants (Cárdenas-[67]). Infection and suboptimal fit were found as the main complications of PEEK implants in recent studies [68]. Specific surface modifications such as coating (with hydroxyapatite, titanium) or increasing porosity were studied in order to increase osseointegration of PEEK implants however, characteristics of modified PEEK implants have not been studied in craniofacial reconstruction [69].

4.2.7 Biodegradable Polymers

Biodegradable polymers undergo chemical reactions such as hydrolysis or oxidation in the host tissues [70]. Following the biodegradation, some breakdown products are formed. Both the implant and its breakdown products need to be biocompatible. Main biodegradable polymers are α-hydroxy acids such as poly(lactic acid) (PLA), poly(glycolic acid) (PGA), or their copolymer poly(lactic-co-glycolic acid) (PLGA) [71]. Following the biodegradation of these polymers, lactic and glycolic acid are formed. These polymers and polymers containing lactic and glycolytic acid with different ratios are used as suture materials, meshes, plate and screws [71–73].

4.3 Ceramics

Ceramics are made of inorganic molecules and have a crystalline structure [74]. Ceramic implants cause low foreign body response, are resistant to bacterial colonization, can resist high compression, and due to their porous form ceramic implants enable tissue ingrowth [44]. Their main disadvantage is ceramic implants are weak and brittle, they can easily be fractured under tensile, torsional and bending forces [75].

The most common ceramics used in plastic surgery is calcium phosphates [76]. Mainly used calcium phosphates are hydroxyapatite and tricalcium phosphate [77]. They are shown to have osteoinductive and osteoconductive properties in-vitro [78]. These implants are used in reconstruction of non-load-bearing bones of craniofacial skeleton [79].

5 Tissue Engineering and Regenerative Medicine

Tissue engineering is a multidisciplinary field that combines principles of engineering and life sciences to create functional biological tissues [80]. Tissue engineering uses combination of cells, biomaterials, and biochemical factors.

Regenerative medicine is a branch of medical science that focuses on developing methods to replace, repair, or regenerate damaged or diseased tissues [81]. Similar to tissue engineering, it uses stem cells, growth factors, and other biological materials to stimulate the organisms' healing processes [82]. The goal is to restore normal function and structure of tissues that have been compromised due to injury, disease, or aging. Stem cell therapy and tissue engineering are key components of regenerative medicine.

Tissue engineering and regenerative medicine uses each other's products and techniques. While tissue engineering uses stem cells that are produced by regenerative medicine, regenerative medicine uses the scaffolds that are produced by tissue engineering. As tissue engineering and regenerative medicine works together, new implants that have cells and cytokines integrated into their structures are being developed.

5.1 Components of Tissue Engineering

Tissues that are produced by tissue engineering has three main components: cells, matrix, and blood vessels [83].

5.1.1 Cells

Cells are the main source of regeneration in novel engineered tissues. Both differentiated cells and precursor stem cells are used. Stem cells are undifferentiated cells that have potential to develop into different types of cells and tissues of an organism [84]. There are mainly two types of stem cells: embryonic stem cells and adult stem cells [85]. Embryonic stem cells are pluripotent and they have potential to differentiate into all types of cells [86]. However, embryonic stem cells are unstable. Adult stem cells are multipotent and they are involved in tissue repair and regeneration [82]. They are primarily located in the bone marrow. There are two types of adult stem cells are found in the body, hematopoietic stem cells and mesenchymal stem cells. There are also endothelial precursor cells which are thought to derive from hematopoietic stem cells [87]. Mesenchymal stem cells are also found in the adipose tissue and in this case, they are known as adipose derived stem cells [88]. It is now known that there are also specific stem cells found in each tissue [89].

Cytokines, chemokines, growth factors, and metalloproteins have important roles in tissue repair, differentiation and development [90, 91]. Growth factors and pro-inflammatory cytokines, stimulate endothelial cell migration and proliferation [92].

Cells also respond to the changes in the 2D and 3D physical environment. There are cell-to-cell and cell-to-ECM adhesion molecules called integrins, which modulate cell migration, gene expressions, development and differentiation [93].

5.1.2 Matrix

Various types of biomaterials have been used in tissue engineering. Biomaterials that have already been used in clinical practice are discussed above. Biomaterials have different roles in tissue engineering. They provide support, they affect cells with biophysical interactions, and they also can contain and release cytokines [80, 83, 94, 95].

Biomaterials

Biomaterials used in tissue engineering can be natural or synthetic [96]. Polymers, ceramics, hydrogels, decellularized allogeneic or xenogeneic tissues are used.

Natural biomaterials used in tissue engineering are proteins (e.g., collagen), polysaccharides (e.g., chitosan, hyaluronic acid), polynucleotides, and extracts of ECM components [97]. The downside of using natural by materials is that they may cause immunologic responses in the host [82].

Synthetic polymers are used as biomaterials. Polymers like polyesters, poly(glycolic acid), poly(lactic acid), poly(lactide-co-glycolide) and ceramic biomaterials such as hydroxyapatite, that are mentioned above are used in tissue engineering as well as polymers like poly(ε-caprolactone)(PCL) [98, 99].

Ceramic biomaterials are particularly used in engineering for hard tissues such as bone tissue engineering [100]. Ceramics such as hydroxyapatite can be used to cover surface of other implants in order to increase osseointegration [101]. Hydrogels are cross-linked polymers that absorb water. They contain physical and chemical bonds within the structure. There are natural hydrogels such as collagen, gelatin, hyaluronic acid, alginate and synthetic hydrogels such as polymers of poly(ethylene glycol)(PEG) [102].

Polymers that change their 3D molecular structure in response to environmental changes are called smart polymers [103]. Recent studies on tissue engineering are focused on use of smart polymers to release certain drugs or cytokines with certain triggers [104].

Fabrication

Biomaterials that are going to be used in tissue engineering can also be fabricated in different forms. Biomaterials in porous shape, hydrogels, meshes, microspheres can be fabricated [105, 106]. There are many techniques to fabricate bio materials: solid free-form fabrication, rapid prototyping methods, polymer phase separation, solid, freeform fabrication, rapid, prototyping methods, polymer phase separation, particle or foam templating, electrospinning techniques have been used [44, 82].

Release Systems

Molecules that have an effect on cell behavior such as growth factors or cytokines can be added into biomaterials structure. Biomaterials can be designed to release these bioactive molecules. Release patterns can be designed to be continuous or pulsatile [107–109]. Desired bioactive molecule can be encapsulated by a biodegradable polymer, which can be attached to the surface of the biomaterial [110]. Furthermore, not only bioactive molecules, but also cells can be designed to be released from the implant [111].

5.1.3 Blood Vessels

Vascular supply is the third main component of an engineered tissue. Mainly implants are dependent to the circulation of the recipient site and they are supplied with diffusion. In order to increase vascularization, biomaterials are loaded with angiogenic cytokines or growth factors. Other methods to increase vascularization of biomaterials are [112–114]:

- cell seeding: incorporating implants with endothelial cells that promote angiogenesis.
- microfabrication: creating micro-tunnels in the implant so that new vessels can generate through the implant.
- co-culture systems: co-culturing enables combination of endothelial cells and structural supportive cells in the implant.
- use of biodegradable scaffolds: biodegradation of the implant enables natural blood vessel formation in degraded areas of the implant.
- pre-vascularization: establishing a blood network within the implant in-vitro or in-vivo prior to the implantation.

5.2 Bone Tissue Engineering

Ideal bone tissue engineering materials need to be osteoconductive, meaning that the implant should support the ingrowth and migration of bone progenitor cells [96]. Some ceramic such as hydroxyapatite, and polymer implants, are known to be osteoconductive. PLA and PCL implants have been studied as bone substitutes. PCL based implants are clinically used.

Another quality of an ideal bone implant is osteoinduction. Biomaterials with scaffolds containing silica and calcium, sodium and phosphorous oxides as well as specifically designed delivery systems increase bone formation in the host [115].

An ideal engineered bone tissue needs to have a porous structure for transportation and angiogenesis, with pores > 300 μm [116].

Strength of a bone implant is of crucial importance. The implant needs to be strong enough to support load bearing host bone, however, needs to avoid “stress shielding” which is caused when implant bears all the tension on the bone that a secondary osteopenia is formed.

Osteoinductive properties of the implant can be increased by incorporation of growth factors and bone morphogenic proteins (BMPs) to the biomaterial [117]. Growth factors such as TGF-B1, FGF-2, platelet-derived growth factors can be combined [118–120]. Bone forming cells such as osteoprogenitor cells or stem cells can also be added to the implant [115, 120–122].

6 3D Printing and Bio-ink

With advances in 3D printing technology, synthetic engineered bone grafts that are specifically "printed" for a craniofacial bone defect can be made [123]. Biomaterials with desired pore sizes can be printed [124]. 3D printing technology can also help to increase osseointegration properties of biomaterials [125]. However, degradation rates and biocompatibility of 3D printed bone implants still need to be improved [126, 127].

A novel alternative method in biomaterial fabrication is to use multicellular aggregates to form a bio-ink, which was printed to form 3D structures [128]. With the integration of release systems and stimuli responsive biomaterials, 3D printing has evolved to 4D printing [129].

7 Future Aspects

An ideal bone implant is yet to be found. Researchers are improving biomaterials in order to increase osteoconductive and osteoinductive properties as well as to enhance bone regeneration and integration. Advances in tissue engineering and regenerative medicine, 3D printing technology contribute to the improvements in bone implants [130, 131].

8 Conclusion

Craniofacial bone defects can occur as a result of congenital defects, traumas, or tumor resections. The gold standard treatment of cranial facial bone defects is the use of autologous bone grafts. Alloplastic reconstruction methods of cranial facial bone defects have been defined in cases where autologous reconstruction is not available. In cases, where neither autologous bone grafts nor allogenic materials are applicable, implants can be used. Commonly used implant materials are metals, polymers, and ceramics. With the advances in tissue engineering and regenerative medicine to enhance healing, to increase bone formation and regeneration, biomaterials that are integrated with progenitor cells, cytokines, and growth factors are developed. An ideal alloplastic bone implant has not been found yet.

References

1. Alkhaibary A, Alharbi A, Alnefaie N, Oqalaa Almubarak A, Aloraidi A, Khairy S (2020) Cranioplasty: a comprehensive review of the history, materials, surgical aspects, and complications. World Neurosurg 139:445–452

2. Oliver JD, Banuelos J, Abu-Ghname A, Vyas KS, Sharaf B (2019) Alloplastic cranioplasty reconstruction: a systematic review comparing outcomes with titanium mesh, polymethyl methacrylate, polyether ether ketone, and norian implants in 3591 adult patients. Ann Plast Surg 82(5S Suppl 4):S289-s294
3. Peng Y, Wang J, Dai X, Chen M, Bao Z, Yang X, Xie J, Wang C, Shao J, Han H et al (2022) Precisely tuning the pore-wall surface composition of bioceramic scaffolds facilitates angiogenesis and orbital bone defect repair. ACS Appl Mater Interfaces 14(38):43987–44001
4. Sasaki R, Togashi C, Miyamoto NS, Okamoto T (2022) Airway compromise in mandibular fracture. J Craniofac Surg 33(4):e385–e386
5. Walshaw EG, Taylor R, Anderson J, Sexton P, Parmar JD, Carter LM (2022) The psychological sequelae of maxillofacial trauma: a scoping review of the literature. Br J Oral Maxillofac Surg 60(10):1303–1320
6. Hurley CM, McConn Walsh R, Shine NP, O'Neill JP, Martin F, O'Sullivan JB (2023) Current trends in craniofacial reconstruction. Surgeon 21(3):e118–e125
7. Christenson E, Figy S, Upadhyay U, Tracy J (2021) Syndrome of the trephined: a rare challenge in head & neck reconstruction. Am J Otolaryngol 42(1):102781
8. Mustroph CM, Stewart CM, Mann LM, Saberian S, Deibert CP, Thompson PW (2022) Systematic review of syndrome of the trephined and reconstructive implications. J Craniofac Surg 33(6):e647–e652
9. Sveikata L, Vasung L, El Rahal A, Bartoli A, Bretzner M, Schaller K, Schnider A, Leemann B (2022) Syndrome of the trephined: Clinical spectrum, risk factors, and impact of cranioplasty on neurologic recovery in a prospective cohort. Neurosurg Rev 45(2):1431–1443
10. Mericli AFSJCHMM (2020) Reconstruction of the scalp, forehead, calvarium, skull base, and midface. In: KC C (ed) Grabb and smith's plastic surgery. Eight Edition ed. Wolters Kluwer
11. De Cola MC, Corallo F, Pria D, Lo Buono V, Calabrò RS (2018) Timing for cranioplasty to improve neurological outcome: a systematic review. Brain Behav 8(11):e01106
12. Xu H, Niu C, Fu X, Ding W, Ling S, Jiang X, Ji Y (2015) Early cranioplasty vs. Late cranioplasty for the treatment of cranial defect: a systematic review. Clin Neurol Neurosurg 136:33–40
13. Zins JE, Langevin CJ, Nasir S (2010) Controversies in skull reconstruction. J Craniofac Surg 21(6):1755–1760
14. Morgan RD, Kharbat AF, Collins RA, Garza J, Belirgen M, Nagy L (2023) Analysis of the timing and the usage of drains following cranioplasty on outcomes and the incidence of bone resorption. Surg Neurol Int 14:329
15. Malcolm JG, Rindler RS, Chu JK, Chokshi F, Grossberg JA, Pradilla G, Ahmad FU (2018) Early cranioplasty is associated with greater neurological improvement: a systematic review and meta-analysis. Neurosurgery 82(3):278–288
16. Oberoi MK, Mirzaie S, Huang KX, Caprini RM, Hu VJ, Dejam D, Ge S, Cronin BJ, Pfaff MJ, Lee JC (2023) Meta-analysis and meta-regression of complications and failures of autologous heterotopic cranial bone versus alloplastic cranioplasties. Plast Reconstr Surg
17. Still MEH, Samant S, Alvarado A, Neal D, Governale LS, Ching JA (2023) Considerations for choice of cranioplasty material for pediatric patients. Pediatr Neurosurg 58(1):1–7
18. Di Rienzo A, Colasanti R, Dobran M, Carrassi E, Herber N, Paracino R, Raggi A, Iacoangeli M (2023) Bone flap resorption after cranioplasty: Risk factors and proposal of the flap integrity score. World Neurosurg
19. Donnelly BM, Smolar DE, Baig AA, Soliman MAR, Monteiro A, Gibbons KJ, Levy EI, Snyder KV (2023) Analysis of craniectomy bone flaps stored in a neurosurgical cryopreservation freezer: microorganism culture results and reimplantation rates. Acta Neurochir (Wien) 165(11):3187–3195
20. Ye J, Liang M, Qiu Q, Zhang W, Ye M (2023) Decompressive bone flap replacement (decompressive cranioplasty): a novel technique for intracranial hypertension-initial experience and outcome. J Neurol Surg A Cent Eur Neurosurg
21. Burke RMM, RJ, Wolfe SA (2013) Acquired cranial and facial bone deformities. In: PC N (ed) Plastic surgery. Elsevier, p 226–242

22. Cabbad NC, Stalder MW, Arroyave A, Wolfe EM, Wolfe SA (2019) Autogenous bone cranioplasty: review of a 42-year experience by a single surgeon. Plast Reconstr Surg 143(6):1713–1723
23. Schmidt AH (2021) Autologous bone graft: Is it still the gold standard? Injury 52(Suppl 2):S18-s22
24. Liu W, Yuan Y, Xiong N, Wang Q, Zhang F, Zhao H, Xu H, Nayaz A, Hendrik P, Sean DJ (2021) Reconstruction of craniectomy for microvascular decompression with autologous particulate bone. J Neurol Surg A Cent Eur Neurosurg 82(6):538–543
25. Stahl A, Yang YP (2021) Regenerative approaches for the treatment of large bone defects. Tissue Eng Part B Rev 27(6):539–547
26. Brink O (2021) The choice between allograft or demineralized bone matrix is not unambiguous in trauma surgery. Injury 52(Suppl 2):S23-s28
27. Gruskin E, Doll BA, Futrell FW, Schmitz JP, Hollinger JO (2012) Demineralized bone matrix in bone repair: history and use. Adv Drug Deliv Rev 64(12):1063–1077
28. Boyce T, Edwards J, Scarborough N (1999) Allograft bone. The influence of processing on safety and performance. Orthop Clin North Am 30(4):571–581
29. Kattz J (2010) The effects of various cleaning and sterilization processes on allograft bone incorporation. J Long Term Eff Med Implants 20(4):271–276
30. Yang P, Zhou J, Ai Q, Yu B, Deng M, Luo F, Xie Z, Xing J, Hou T (2020) Comparison of individual tissue-engineered bones and allogeneic bone in treating bone defects: a long-term follow-up study. Cell Transplant 29:963689720940722
31. Lomas R, Chandrasekar A, Board TN (2013) Bone allograft in the u.K.: perceptions and realities. Hip Int 23(5):427–433
32. Urist MR (1965) Bone: formation by autoinduction. Science 150(3698):893–899
33. Vargel I, Tunçbilek G, Mavili E, Cila A, Ruacan S, Benli K, Erk Y (2004) Solvent-dehydrated calvarial allografts in craniofacial surgery. Plast Reconstr Surg 114(2):298–306
34. Guler S, Eichholz K, Chariyev-Prinz F, Pitacco P, Aydin HM, Kelly DJ, Vargel İ (2022) Biofabrication of poly(glycerol sebacate) scaffolds functionalized with a decellularized bone extracellular matrix for bone tissue engineering. Bioengineering (Basel) 10(1)
35. Zhang H, Yang L, Yang XG, Wang F, Feng JT, Hua KC, Li Q, Hu YC (2019) Demineralized bone matrix carriers and their clinical applications: an overview. Orthop Surg 11(5):725–737
36. Amini Z, Lari R (2021) A systematic review of decellularized allograft and xenograft-derived scaffolds in bone tissue regeneration. Tissue Cell 69:101494
37. Canter HI, Vargel I, Mavili ME (2007) Reconstruction of mandibular defects using autografts combined with demineralized bone matrix and cancellous allograft. J Craniofac Surg 18(1):95–100; discussion 101–103
38. Bracey Daniel N, Jinnah Alexander H, Willey Jeffrey S, Seyler Thorsten M, Hutchinson Ian D, Whitlock Patrick W, Smith Thomas L, Danelson Kerry A, Emory Cynthia L, Kerr BA (2019) Investigating the osteoinductive potential of a decellularized xenograft bone substitute. Cells Tissues Organs 207(2):97–113
39. Bracey DN, Seyler TM, Jinnah AH, Lively MO, Willey JS, Smith TL, Van Dyke ME, Whitlock PW (2018) A decellularized porcine xenograft-derived bone scaffold for clinical use as a bone graft substitute: a critical evaluation of processing and structure. J Funct Biomater 9(3):45
40. Bracey DN, Seyler TM, Jinnah AH, Smith TL, Ornelles DA, Deora R, Parks GD, Van Dyke ME, Whitlock PW (2019) A porcine xenograft-derived bone scaffold is a biocompatible bone graft substitute: an assessment of cytocompatibility and the alpha-gal epitope. Xenotransplantation 26(5):e12534
41. Couso-Queiruga E, Weber HA, Garaicoa-Pazmino C, Barwacz C, Kalleme M, Galindo-Moreno P, Avila-Ortiz G (2023) Influence of healing time on the outcomes of alveolar ridge preservation using a collagenated bovine bone xenograft: a randomized clinical trial. J Clin Periodontol 50(2):132–146
42. Karalashvili L, Kakabadze A, Uhryn M, Vyshnevska H, Ediberidze K, Kakabadze Z (2018) Bone grafts for reconstruction of bone defects (review). Georgian Med News 282:44–49

43. Galindo-Moreno P, Abril-García D, Carrillo-Galvez AB, Zurita F, Martín-Morales N, O'Valle F, Padial-Molina M (2022) Maxillary sinus floor augmentation comparing bovine versus porcine bone xenografts mixed with autogenous bone graft. A split-mouth randomized controlled trial. Clin Oral Implants Res 33(5):524–536
44. Butler CEK, TW (2013) Implants and biomaterials. In: PC N (ed) Plastic surgery. Elsevier, p 786–797
45. Cumberland VH (1952) A preliminary report on the use of prefabricated nylon weave in the repair of ventral hernia. Med J Aust 1(5):143–144
46. JT S (1953) Materials for hernia repair. Proc R Soc Med (46):647–652
47. Fan L, Chen S, Yang M, Liu Y, Liu J (2023) Metallic materials for bone repair. Adv Healthc Mater e2302132
48. Pacifici L, F DEA, Orefici A, Cielo A (2016) Metals used in maxillofacial surgery. Oral Implantol (Rome) 9(Suppl 1/2016 to N 4/2016):107–111
49. Disegi JA, Eschbach L (2000) Stainless steel in bone surgery. Injury 31(Suppl 4):2–6
50. Turner S, Harrison JE, Sharif FN, Owens D, Millett DT (2021) Orthodontic treatment for crowded teeth in children. Cochrane Database Syst Rev 12(12):Cd003453
51. Barone CM, Eisig S, Wallach S, Mitnick R, Mednick R (1994) Effects of rigid fixation device composition on three-dimensional computed axial tomography imaging: direct measurements on a pig model. J Oral Maxillofac Surg 52(7):737–740; discussion 740–731
52. Lim HK, Choi YJ, Choi WC, Song IS, Lee UL (2022) Reconstruction of maxillofacial bone defects using patient-specific long-lasting titanium implants. Sci Rep 12(1):7538
53. Anele CO, Balogun SA, Ezeaku CO, Ajekwu TO, Omon HE, Ejembi GO, Komolafe EO (2024) Titanium mesh cranioplasty for cosmetically disfiguring cranio-facial tumours in a resource limited setting. World Neurosurg X 23:100362
54. Umair M, Javed I, Rehman M, Madni A, Javeed A, Ghafoor A, Ashraf M (2016) Nanotoxicity of inert materials: the case of gold, silver and iron. J Pharm Pharm Sci 19(2):161–180
55. Şahin MM, Uzunoğlu E, Karamert R, Cebeci S, Cesur G, Yalçın M, Düzlü M, Tutar H, Uğur MB, Ceylan A (2021) The role of gold weight implants in the management of paralytic lagophthalmos. Turk J Med Sci 51(5):2584–2591
56. Oh TS, Min K, Song SY, Choi JW, Koh KS (2018) Upper eyelid platinum weight placement for the treatment of paralytic lagophthalmos: a new plane between the inner septum and the levator aponeurosis. Arch Plast Surg 45(3):222–228
57. Li S, Zhang J, He J, Liu W, Wang Y, Huang Z, Pang H, Chen Y (2023) Functional pdms elastomers: bulk composites, surface engineering, and precision fabrication. Adv Sci (Weinh) e2304506
58. Mwanza JC, Ngoy DK, Kayembe DL (2001) Reconstruction of orbital floor blow-out fractures with silicone implant. Bull Soc Belge Ophtalmol 280:57–61
59. Urban IA, Saleh MHA, Ravidà A, Forster A, Wang HL, Barath Z (2021) Vertical bone augmentation utilizing a titanium-reinforced ptfe mesh: a multi-variate analysis of influencing factors. Clin Oral Implants Res 32(7):828–839
60. Rosen MJ (2009) Polyester-based mesh for ventral hernia repair: is it safe? Am J Surg 197(3):353–359
61. Seifalian A, Basma Z, Digesu A, Khullar V (2023) Polypropylene pelvic mesh: what went wrong and what will be of the future? Biomedicines 11(3)
62. de Moraes Ferreira AC, Muñoz XM, Okamoto R, Pellizer EP, Garcia IR Jr (2016) Postoperative complications in craniomaxillofacial reconstruction with medpor. J Craniofac Surg 27(2):425–428
63. Chen K, Schultz BD, Mattos D, Reish RG (2022) Optimizing the use of autografts, allografts, and alloplastic materials in rhinoplasty. Plast Reconstr Surg 150(3):675e–683e
64. Perozzo FAG, Ku YC, Kshettry VR, Sikder P, Papay FA, Rampazzo A, Bassiri Gharb B (2024) High-density porous polyethylene implant cranioplasty: a systematic review of outcomes. J Craniofac Surg
65. Ma R, Tang T (2014) Current strategies to improve the bioactivity of peek. Int J Mol Sci 15(4):5426–5445

66. Scolozzi P, Martinez A, Jaques B (2007) Complex orbito-fronto-temporal reconstruction using computer-designed peek implant. J Craniofac Surg 18(1):224–228
67. Cárdenas-Serres C, Almeida-Parra F, Simón-Flors AM, de Leyva-Moreno P, Ranz-Colio Á, Ley-Urzaiz L, Acero-Sanz J (2024) Custom cad/cam peek implants for complex orbitocranial reconstruction: our experience with 15 patients. J Clin Med 13(3)
68. Shewmake C, Khetpal S, Ozaki W (2023) Polyether ether ketone (peek) implants and the manufacturer and user facility device experience (maude) database: an updated review of adverse events and complications. J Craniofac Surg
69. Kauke-Navarro M, Knoedler L, Knoedler S, Deniz C, Safi AF (2024) Surface modification of peek implants for craniofacial reconstruction and aesthetic augmentation-fiction or reality? Front Surg 11:1351749
70. Bher A, Cho Y, Auras R (2023) Boosting degradation of biodegradable polymers. Macromol Rapid Commun 44(5):e2200769
71. Oleksy M, Dynarowicz K, Aebisher D (2023) Advances in biodegradable polymers and biomaterials for medical applications-a review. Molecules 28(17)
72. Choi SH, Lee JH (2016) Absorbable plate-related infection after facial bone fracture reduction. Arch Craniofac Surg 17(1):1–4
73. Koutserimpas C, Alpantaki K, Chatzinikolaidou M, Chlouverakis G, Dohm M, Hadjipavlou AG (2018) The effectiveness of biodegradable instrumentation in the treatment of spinal fractures. Injury 49(12):2111–2120
74. Mantripragada VP, Lecka-Czernik B, Ebraheim NA, Jayasuriya AC (2013) An overview of recent advances in designing orthopedic and craniofacial implants. J Biomed Mater Res A 101(11):3349–3364
75. Traina F, De Fine M, Di Martino A, Faldini C (2013) Fracture of ceramic bearing surfaces following total hip replacement: a systematic review. Biomed Res Int 2013:157247
76. Thrivikraman G, Athirasala A, Twohig C, Boda SK, Bertassoni LE (2017) Biomaterials for craniofacial bone regeneration. Dent Clin North Am 61(4):835–856
77. Suzuki O, Shiwaku Y, Hamai R (2020) Octacalcium phosphate bone substitute materials: comparison between properties of biomaterials and other calcium phosphate materials. Dent Mater J 39(2):187–199
78. Le Guéhennec L, Soueidan A, Layrolle P, Amouriq Y (2007) Surface treatments of titanium dental implants for rapid osseointegration. Dent Mater 23(7):844–854
79. De Benedictis A, Sawant N, Marasi A, Rossi-Espagnet MC, Carai A, Luglietto D, Bua A, Randi F, Savioli A, Borro L et al (2023) Calcium phosphate and titanium cranioplasty after total angular craniopagus separation. Clin Neurol Neurosurg 232:107906
80. Zhang J, Xu W, Li C, Meng F, Guan Y, Liu X, Zhao J, Peng J, Wang Y (2022) Tissue engineering microtissue: construction, optimization, and application. Tissue Eng Part B Rev 28(2):393–404
81. Edgar L, Pu T, Porter B, Aziz JM, La Pointe C, Asthana A, Orlando G (2020) Regenerative medicine, organ bioengineering and transplantation. Br J Surg 107(7):793–800
82. O'Connor AJM (2013) Tissue engineering. In: Neligan PC (ed) Plastic surgery. Elsevier, pp 367–396
83. Guan Y, Yang B, Xu W, Li D, Wang S, Ren Z, Zhang J, Zhang T, Liu X, Li J et al (2022) Cell-derived extracellular matrix materials for tissue engineering. Tissue Eng Part B Rev 28(5):1007–1021
84. Zhang Z, Gupte MJ, Ma PX (2013) Biomaterials and stem cells for tissue engineering. Expert Opin Biol Ther 13(4):527–540
85. Zakrzewski W, Dobrzyński M, Szymonowicz M, Rybak Z (2019) Stem cells: past, present, and future. Stem Cell Res Ther 10(1):68
86. Reynolds J, Lamba DA (2014) Human embryonic stem cell applications for retinal degenerations. Exp Eye Res 123:151–160
87. Peters EB (2018) Endothelial progenitor cells for the vascularization of engineered tissues. Tissue Eng Part B Rev 24(1):1–24

88. Mazini L, Rochette L, Admou B, Amal S, Malka G (2020) Hopes and limits of adipose-derived stem cells (adscs) and mesenchymal stem cells (mscs) in wound healing. Int J Mol Sci 21(4)
89. Bacakova L, Zarubova J, Travnickova M, Musilkova J, Pajorova J, Slepicka P, Kasalkova NS, Svorcik V, Kolska Z, Motarjemi H et al (2018) Stem cells: their source, potency and use in regenerative therapies with focus on adipose-derived stem cells—a review. Biotechnol Adv 36(4):1111–1126
90. Amarasekara DS, Kim S, Rho J (2021) Regulation of osteoblast differentiation by cytokine networks. Int J Mol Sci 22(6)
91. Hu C, Chan SI, Sawyer EB, Yu Y, Wang J (2014) Metalloprotein design using genetic code expansion. Chem Soc Rev 43(18):6498–6510
92. Li X, Kumar A, Carmeliet P (2019) Metabolic pathways fueling the endothelial cell drive. Annu Rev Physiol 81:483–503
93. Hight-Warburton W, Parsons M (2019) Regulation of cell migration by α4 and α9 integrins. Biochem J 476(4):705–718
94. Costa-Pinto AR, Vargel I, Tuzlakoglu K, Correlo VM, Sol PC, Faria S, Piskin E, Reis RL, Neves NM (2014) Influence of scaffold composition over in vitro osteogenic differentiation of hbmscs and in vivo inflammatory response. J Biomater Appl 28(9):1430–1442
95. Moutos FT, Estes BT, Guilak F (2010) Multifunctional hybrid three-dimensionally woven scaffolds for cartilage tissue engineering. Macromol Biosci 10(11):1355–1364
96. Perić Kačarević Ž, Rider P, Alkildani S, Retnasingh S, Pejakić M, Schnettler R, Gosau M, Smeets R, Jung O, Barbeck M (2020) An introduction to bone tissue engineering. Int J Artif Organs 43(2):69–86
97. Meyer M (2019) Processing of collagen based biomaterials and the resulting materials properties. Biomed Eng Online 18(1):24
98. Ekin O, Calis M, Aliyev A, Yar AS, Korkusuz P, Bilgic E, Aydin HM, Celik HH, Ozgur F, Vargel I (2016) Poly(l-lactide)/poly(ε-caprolactone) and collagen/β-tricalcium phosphate scaffolds for the treatment of critical-sized rat alveolar defects: a microtomographic, molecular-biological, and histological study. Cleft Palate Craniofac J 53(4):453–463
99. Siddiqui N, Kishori B, Rao S, Anjum M, Hemanth V, Das S, Jabbari E (2021) Electropsun polycaprolactone fibres in bone tissue engineering: a review. Mol Biotechnol 63(5):363–388
100. Denry I, Kuhn LT (2016) Design and characterization of calcium phosphate ceramic scaffolds for bone tissue engineering. Dent Mater 32(1):43–53
101. Sirin HT, Vargel I, Kutsal T, Korkusuz P, Piskin E (2016) Ti implants with nanostructured and ha-coated surfaces for improved osseointegration. Artif Cells Nanomed Biotechnol 44(3):1023–1030
102. Atwal A, Dale TP, Snow M, Forsyth NR, Davoodi P (2023) Injectable hydrogels: an emerging therapeutic strategy for cartilage regeneration. Adv Colloid Interface Sci 321:103030
103. Khan HM, Liao X, Sheikh BA, Wang Y, Su Z, Guo C, Li Z, Zhou C, Cen Y, Kong Q (2022) Smart biomaterials and their potential applications in tissue engineering. J Mater Chem B 10(36):6859–6895
104. Huang HJ, Tsai YL, Lin SH, Hsu SH (2019) Smart polymers for cell therapy and precision medicine. J Biomed Sci 26(1):73
105. Hu J, Albadawi H, Chong BW, Deipolyi AR, Sheth RA, Khademhosseini A, Oklu R (2019) Advances in biomaterials and technologies for vascular embolization. Adv Mater 31(33):e1901071
106. Zeimaran E, Pourshahrestani S, Fathi A, Razak N, Kadri NA, Sheikhi A, Baino F (2021) Advances in bioactive glass-containing injectable hydrogel biomaterials for tissue regeneration. Acta Biomater 136:1–36
107. Canter HI, Vargel I, Korkusuz P, Oner F, Gungorduk DB, Cil B, Karabulut E, Sargon MF, Erk Y (2010) Effect of use of slow release of bone morphogenetic protein-2 and transforming growth factor-beta-2 in a chitosan gel matrix on cranial bone graft survival in experimental cranial critical size defect model. Ann Plast Surg 64(3):342–350

108. Pişkin E, Işoğlu IA, Bölgen N, Vargel I, Griffiths S, Cavuşoğlu T, Korkusuz P, Güzel E, Cartmell S (2009) In vivo performance of simvastatin-loaded electrospun spiral-wound polycaprolactone scaffolds in reconstruction of cranial bone defects in the rat model. J Biomed Mater Res A 90(4):1137–1151
109. Zhang J, Zheng Y, Lee J, Hua J, Li S, Panchamukhi A, Yue J, Gou X, Xia Z, Zhu L et al (2021) A pulsatile release platform based on photo-induced imine-crosslinking hydrogel promotes scarless wound healing. Nat Commun 12(1):1670
110. Zhai X, Wu Y, Tan H (2023) Gelatin-based targeted delivery systems for tissue engineering. Curr Drug Targets 24(8):673–687
111. Xu M, Qin M, Cheng Y, Niu X, Kong J, Zhang X, Huang D, Wang H (2021) Alginate microgels as delivery vehicles for cell-based therapies in tissue engineering and regenerative medicine. Carbohydr Polym 266:118128
112. Moore MJ, Tan RP, Yang N, Rnjak-Kovacina J, Wise SG (2022) Bioengineering artificial blood vessels from natural materials. Trends Biotechnol 40(6):693–707
113. Newman Frisch A, Debbi L, Shuhmaher M, Guo S, Levenberg S (2022) Advances in vascularization and innervation of constructs for neural tissue engineering. Curr Opin Biotechnol 73:188–197
114. Simunovic F, Finkenzeller G (2021) Vascularization strategies in bone tissue engineering. Cells. 10(7)
115. Li W, Liu Y, Zhang P, Tang Y, Zhou M, Jiang W, Zhang X, Wu G, Zhou Y (2018) Tissue-engineered bone immobilized with human adipose stem cells-derived exosomes promotes bone regeneration. ACS Appl Mater Interfaces 10(6):5240–5254
116. Karageorgiou V, Kaplan D (2005) Porosity of 3d biomaterial scaffolds and osteogenesis. Biomaterials 26(27):5474–5491
117. Torres-Guzman RA, Avila FR, Maita KC, Garcia JP, De Sario GD, Borna S, Eldaly AS, Quinones-Hinojosa A, Zubair AC, Ho OA et al (2023) Bone morphogenic protein and mesenchymal stem cells to regenerate bone in calvarial defects: a systematic review. J Clin Med. 12(12)
118. Alici-Garipcan A, Korkusuz P, Bilgic E, Askin K, Aydin HM, Ozturk E, Inci I, Ozkizilcik A, Kamile Ozturk K, Piskin E et al (2019) Critical-size alveolar defect treatment via tgf-ß3 and bmp-2 releasing hybrid constructs. J Biomater Sci Polym Ed 30(5):415–436
119. Oliveira ÉR, Nie L, Podstawczyk D, Allahbakhsh A, Ratnayake J, Brasil DL, Shavandi A (2021) Advances in growth factor delivery for bone tissue engineering. Int J Mol Sci 22(2)
120. Vural AC, Odabas S, Korkusuz P, Yar Sağlam AS, Bilgiç E, Çavuşoğlu T, Piskin E, Vargel İ (2017) Cranial bone regeneration via bmp-2 encoding mesenchymal stem cells. Artif Cells Nanomed Biotechnol 45(3):544–550
121. Bölgen N, Korkusuz P, Vargel İ, Kılıç E, Güzel E, Çavuşoğlu T, Uçkan D, Pişkin E (2014) Stem cell suspension injected hema-lactate-dextran cryogels for regeneration of critical sized bone defects. Artif Cells Nanomed Biotechnol 42(1):70–77
122. İşoğlu İA, Bölgen N, Korkusuz P, Vargel İ, Çelik HH, Kılıç E, Güzel E, Çavuşoğlu T, Uçkan D, Pişkin E (2019) Stem cells combined 3d electrospun nanofibrous and macrochannelled matrices: a preliminary approach in repair of rat cranial bones. Artif Cells Nanomed Biotechnol 47(1):1094–1100
123. Su X, Wang T, Guo S (2021) Applications of 3d printed bone tissue engineering scaffolds in the stem cell field. Regen Ther 16:63–72
124. Distler T, Fournier N, Grünewald A, Polley C, Seitz H, Detsch R, Boccaccini AR (2020) Polymer-bioactive glass composite filaments for 3d scaffold manufacturing by fused deposition modeling: fabrication and characterization. Front Bioeng Biotechnol 8
125. Tian L, Zhang Z, Tian B, Zhang X, Wang N (2020) Study on antibacterial properties and cytocompatibility of epl coated 3d printed pcl/ha composite scaffolds. RSC Adv 10(8):4805–4816
126. Erickson CB, Newsom JP, Fletcher NA, Feuer ZM, Yu Y, Rodriguez-Fontan F, Hadley Miller N, Krebs MD, Payne KA (2020) In vivo degradation rate of alginate-chitosan hydrogels influences tissue repair following physeal injury. J Biomed Mater Res B Appl Biomater 108(6):2484–2494

127. Kazimierczak P, Benko A, Nocun M, Przekora A (2019) Novel chitosan/agarose/hydroxyapatite nanocomposite scaffold for bone tissue engineering applications: Comprehensive evaluation of biocompatibility and osteoinductivity with the use of osteoblasts and mesenchymal stem cells. Int J Nanomed 14(null):6615–6630
128. Badhe RV, Chatterjee A, Bijukumar D, Mathew MT (2023) Current advancements in bio-ink technology for cartilage and bone tissue engineering. Bone 171:116746
129. Wan Z, Zhang P, Liu Y, Lv L, Zhou Y (2020) Four-dimensional bioprinting: current developments and applications in bone tissue engineering. Acta Biomater 101:26–42
130. Pérez-Davila S, Potel-Alvarellos C, Carballo R, González-Rodríguez L, López-Álvarez M, Serra J, Díaz-Rodríguez P, Landin M, González P (2023) Vancomycin-loaded 3d-printed polylactic acid-hydroxyapatite scaffolds for bone tissue engineering. Polymers (Basel) 15(21)
131. Singh I, Dixit K, Gupta P, George SM, Sinha N, Balani K (2023) 3d-printed multifunctional ag/ceo(2)/zno reinforced hydroxyapatite-based scaffolds with effective antibacterial and mechanical properties. ACS Appl Bio Mater

Alaz Çırak , MD (https://avesis.hacettepe.edu.tr/alaz.cirak), is currently a resident at the Department of Plastic Reconstructive and Aesthetic Surgery, Hacettepe University Faculty of Medicine in Turkey. He completed his M.D. at Hacettepe University Faculty of Medicine in 2020. His professional interests lie in plastic surgery, with a specific focus on hand surgery, cleft lip and palate, craniofacial surgery and reconstructive microsurgery. Dr. Çırak's clinical and academic growth has been further enhanced by several international observerships, including one at Washington University's Center for Nerve Injury and Paralysis in St. Louis, Missouri, in 2023. Dr. Çırak has authored and co-authored several articles, including a chapter on the application of botulinum toxin in temporomandibular joint disorders, which was published in 2023. His research article on pharyngeal flap outcomes for velopharyngeal insufficiency was published in the *Journal of Plastic, Reconstructive & Aesthetic Surgery* in 2024, and his study on abdominal wall flaps appeared in *Acta Medica* in 2024. He has also contributed chapters on craniosynostosis, distraction osteogenesis, craniofacial clefts, hypertelorism, and craniofacial microstomia in a national book *Plastik Cerrahi Akıl Notları* published in 2024.

İbrahim Vargel , MD, PhD, (https://scholar.google.com.tr/citations?user=kTo6w74AAAAJ&hl=enhttps://www.google.com/search?client=safari&rls=en&q=ibrahim+vargel+avesis&ie=UTF-8&oe=UTF-8), born in 1955 in Aydın/Nazilli, Turkey, graduated from the Faculty of Medicine at Hacettepe University in 1986. He specialized in Plastic, Reconstructive, and Aesthetic Surgery at Hacettepe University (1990–1997), and further trained in Craniofacial Surgery and Hand Surgery. He earned his PhD in Advanced Plastic Surgery, focusing on Cranio-Maxillo-Facial Surgery, from Hacettepe University (1998–2003). Dr. Vargel expanded his expertise through international experiences, including a clinical fellowship (DAFPS programme) at Erasmus University, Sophia Children's Hospital, Rotterdam (2012), and as a visiting researcher at the University of Pittsburgh (2014), concentrating on Tissue Engineering and Regenerative Medicine. He also completed subspecialty training in Oral and Maxillofacial Surgery and Hand Surgery at the Ministry of Health, Turkey, in 2011 and 2012. Dr. Vargel has served as the president or an executive committee member of numerous societies, including as the president of the Turkish Society of Plastic, Reconstructive, and Aesthetic Surgery. He is currently an active member of EURAPS. He has held prominent academic and clinical roles, serving as faculty at Hacettepe University's Department of Plastic, Reconstructive, and Aesthetic Surgery (1997–2014) and as the Head of the Department at Kırıkkale University (2005–2012). Dr. Vargel is currently a Professor at Hacettepe University and a faculty member in the Biomedical Engineering Department. Dr. Vargel has authored more than 100 international scientific publications, contributed to national publications, and has edited three scientific books. He has supervised 17 doctoral theses and has a total of 3,290 citations according to Google Scholar, with an h-index of 29 and an i10-index of 67. His research interests focus on Craniofacial Anomalies, Vascular Anomalies, Tissue Engineering, and advanced reconstructive aesthetic surgeries.

Exploring the Cutting Edge of Platelet-Rich Plasma Omics

Bilge Başak Fidan, Murat Koçak, İlayda Demirdiş, Hakan Aydınlı, Emine Çiftçi, Ozan Kaplan, Özge Boyacıoğlu, Petek Korkusuz, Mustafa Çelebier, and Feza Korkusuz

Abstract This chapter delves into the landscape of knee osteoarthritis (KOA), a prevalent chronic disorder affecting millions of people worldwide, causing significant morbidity and mortality. The comprehensive chapter examines publications in the Web of Science (WoS) database from 2019 to 2023, focusing on osteoarthritis (OA),

B. B. Fidan (✉) · Ö. Boyacıoğlu
Graduate School of Science and Engineering, Department of Bioengineering, Hacettepe University, Ankara, Türkiye
e-mail: bilgefidan@hacettepe.edu.tr

Ö. Boyacıoğlu
e-mail: ozge.boyacioglu@hacettepe.edu.tr

M. Koçak
Department of Medical Informatics, Faculty of Medicine, Baskent University, Ankara, Türkiye
e-mail: muratkocak@baskent.edu.tr

İ. Demirdiş
Department of Biology and Molecular Biology, Graduate School of Science and Engineering, Hacettepe University, Ankara, Türkiye

H. Aydınlı
Faculty of Medicine, Hacettepe University, Ankara, Türkiye

E. Çiftçi
Department of Sports Medicine Ankara, Faculty of Medicine, Hacettepe University, Ankara, Türkiye

O. Kaplan · M. Çelebier
Department of Analytical Chemistry, Faculty of Pharmacy, Hacettepe University, Ankara, Türkiye
e-mail: ozankaplan@hacettepe.edu.tr

M. Çelebier
e-mail: celebier@hacettepe.edu.tr

Ö. Boyacıoğlu
Department of Medical Biochemistry, Faculty of Medicine, Atılım University, Ankara, Türkiye

P. Korkusuz
Department of Histology and Embryology, Faculty of Medicine, Hacettepe University, Ankara, Türkiye

MEMS Center, Middle East Technical University, Ankara, Türkiye

N. Sağlam et al. (eds.), *Nano-Biomaterials in Tissue Repair and Regeneration*, Tissue Repair and Reconstruction, https://doi.org/10.1007/978-981-96-3808-6_2

platelet-rich plasma (PRP), and omics studies. The scope extends to the metabolomic analyses of stability tests conducted on PRP, a commonly utilized modality in the treatment of OA. Notably, findings suggest that freezing and reusing PRP may not pose a toxicological risk; however, from a physiological perspective, it may never fully substitute fresh PRP. In light of these results, the study recommends the preparation of PRP in a fresh state for optimal efficacy. This nuanced exploration provides valuable insights into the multifaceted realm of OA research and its potential therapeutic interventions.

Keywords Osteoarthritis · Platelet-rich plasma · Metabolomics · Proteomics · Articular joint cartilage

1 Overview of Knee Osteoarthritis: A Common Chronic Disorder Where PRP is Frequently Applied to Modify the Course of the Condition

Osteoarthritis (OA) is a chronic disabling disease with high incidence of morbidity and mortality [1]. Impaired mobility in OA results in the development of comorbidities such as diabetes mellitus and cardiovascular disease. Mortality in OA was mostly related to cardiovascular diseases. The estimated number of symptomatic OA patients worldwide is about 528 million [2] The OA incidence increased about 9.0% during the last 28 years [3]. The prevalence of OA is 16.1% [4], with a higher occurrence in low and middle-income countries [5]. OA is an unavoidable outcome of aging and prevalence of OA is expected to be increased with the aging population [6]. Over 32.5 million of adults are affected by OA in the US [7]. It is more frequent in 60 + years old females. KOA, that is also named as gonarthrosis, at the age of 65 + years has a prevalence of 33.6% and the number increases up to 75% after the age of 75 years [8]. Gonarthrosis was diagnosed in 654.1 million people in the globe and the incidence was 203 individuals out of 10 thousand people [4]. The prevalence and incidence of KOA in the Middle East and Northern Africa was 5,343 and 430.4 individuals in 100,000 people, respectively [9]. Another study in Europe [10] indicated that the incidence of medicinal use is about 46,9% among OA patients and they face problems in working activities. Another study [11] that evaluated centenarians of the Ural-Altay region highlighted the prevalence of OA as 27.7% and only 13.8% were on medicine for their symptoms. Increasing age and better life expectancies of gonarthrotic patients is increasing the burden of OA patients [3]. A study published

P. Korkusuz
e-mail: petek@hacettepe.edu.tr

F. Korkusuz (✉)
Department of Sports Medicine, Faculty of Medicine, Hacettepe University, Ankara, Türkiye
e-mail: feza.korkusuz@hacettepe.edu.tr

in 2014 [12] revealed that OA will increase from 26.6% to 29.5% in 2032. KOA is observed in about 85% of patients. According to the EU OA report (2022 July-Research Report: Osteoarthritis in the EU), the direct and indirect burden of OA increased to 7.2 billion and 4.4 billion Euros. Extensive research is conducted to unravel the genetic and epigenetic aspects of the disease [13]. Excessive physical loads on the knee joint as seen in abnormal working conditions and extreme sports are also linked to the high prevalence and rapid progression of KOA [14].

2 Sport-Related Post-Traumatic Osteoarthritis

The incidence and prevalence of sport-related injuries are increasing [15]. OA was traditionally considered a condition exclusive to the elderly. The number of secondary and early OA is increasing [16] due to the increase in sport-related injuries. Post-traumatic osteoarthritis (PTOA) can arise as a result of joint injuries and is responsible for around 5.5 million cases in the United States each year [17]. The initial magnitude of trauma, especially in contact sports, may cause joint cartilage injury as well as ruptures in ligaments and other cartilaginous tissues. Chronic joint overload on the other hand may accelerate the development of OA. The incidence of OA in the knees and ankles of former high impact athletes has been shown to be higher than in the normal population [18]. Athletes in particular who play sports that involve rapid acceleration with sudden deceleration, such as soccer, volleyball and basketball or who train continuously with high impact on the joints or compete at elite level for long periods of time are more likely to develop OA [19].

Ligament injuries, such as tears in the anterior cruciate ligament (ACL), are believed to cause a reduction in damping and shock absorption capabilities. This leads to increased loads being transferred to other joint structures, including cartilage, when the skeletal muscles are unable to maintain stability. Major knee joint injuries such as ACL rupture and acute meniscal tears are among the major risk factors for KOA and their incidence is increasing [20]. Skeletal muscle weakness is another common consequence of joint injuries and reduced lower limb muscle strength usually plays a significant role in the development and progression of KOA. There is a report indicating that PTOA is different as it occurs in younger people and is faster progressing. Guidelines for preventing and managing PTOA are established on the most reliable evidence. Individuals at risk of PTOA following acute traumatic joint injury should be encouraged to adopt measures such as primary injury prevention, self-management strategies, maintaining a healthy body weight, and engaging in an appropriate level of physical activity. It is crucial to educate athletic trainers and patients about PTOA for effective prevention and the development of management strategies for this condition [21].

Diagnosis and treatment of primary and secondary KOA is multidimensional (Fig. 1). It is age and stage dependent. Understanding the needs and expectations of the patient is very important. Symptoms and laboratory findings may change with the progression of the disease. Inflammatory and biomechanical factors contributes

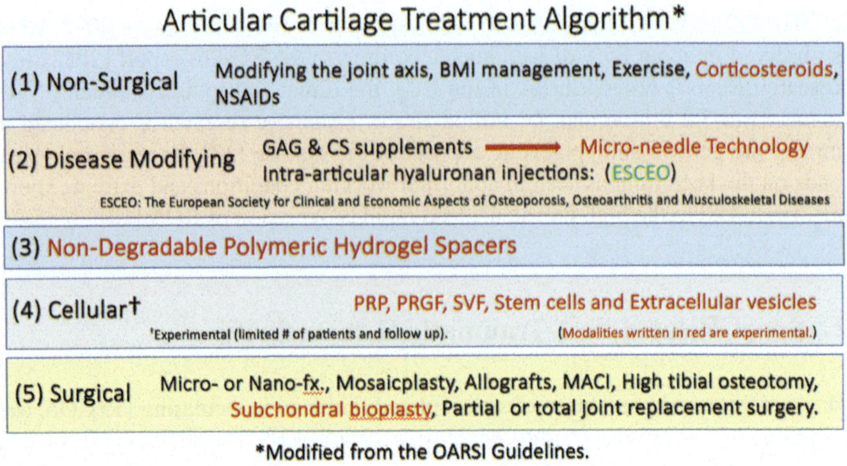

Fig. 1 Knee osteoarthritis treatment algorithm

in the complex pathophysiology of the disease as opposed to the former belief that it was purely a wear and tear disorder [22]. Synovial fluid examination [23] apart from other diagnostic tools such as conventional x-rays and recently ultrasound and/or magnetic resonance imaging is an important but less-used tool for precise diagnosis.

Joint cartilage is avascular and its metabolic activity is balanced through the synovial fluid (SF). Intra-articular injections are therefore quite common compared to other applications. Disease modifying peri- and intra-articular injections also enabled researchers to obtain synovial fluid at various stages of injury and/or disease. Such an approach may also allow the evaluation of remission or cure of the disease at or after intra-articular molecular and/or biological repetitive injections.

3 Platelet-Rich Plasma as a Strong Anti-inflammatory and Regenerative Treatment Modality in Knee Osteoarthritis

3.1 Platelet-Rich Plasma

Platelet-rich plasma (PRP) is a component of PRP obtained as a result of centrifugation of blood. The term PRP was used by hematologists in the 1970s to express that the platelet count is higher in peripheral blood than in plasma [24]. In the preceding decade, it has become obvious that platelets play a significant role in circulation, connecting hemostasis, innate, and acquired immunity. The protein composition, localization, and function are essential for regulating platelet function. The required platelet count in a healthy individual varies between 150,000 and 350,000 cells/μL.

PRP has a 3 to 5 times higher platelet count than the whole blood [24]. Platelets are cytoplasmic fragments of megakaryocytes that differentiate from the myeloid cell line. They do not have a nucleus, therefore they cannot reproduce. Their life span is of approximately seven to ten days in the bone marrow [25, 26].

More than one factor activates platelets. There are two basic activation mechanisms of platelets. These mechanisms include GP Ib-IX-V, GP VI or C-type lectin-like receptor 2 (CLEC-2), membrane glycoproteins expressed specifically in megakaryocytes and platelets with closely related signal transduction pathways [27]. The most basic marker of platelet activation is the increased level of monocyte-platelet aggregates (MPAs) in the circulation [28]. Platelets are primarily associated with hemostasis and coagulation and are activated as a result of an injury, releasing their granules are filled with growth factor [29].

As a result of this release, the stimulation of the inflammatory cascade and healing process initiates tissue repair [29]. One of the reasons for rapid platelet activation is the interaction of platelets with pathogens. Activated platelets collect neutrophils, macrophages and dendritic cells at the site of infection and secrete pro- and anti-inflammatory molecules and several types of platelet-derived extracellular vesicles [28]. For example, Dengue virus, which affects 2.5 billion of the world's population and causes systemic inflammation and increased vascular permeability, triggers leukocyte-platelet activation [28, 30]. Dengue virus activates platelets via CLEC-2 and neutrophils and macrophages via CLEC5A and TLR2, resulting in the release of extracellular vesicles. As a result, neutrophil extracellular traps (NET) and proinflammatory cytokine release are induced [31].

The platelet proteome contains a complex set of data that is difficult to interpret. In some cases, various approaches have been developed to increase the scope of proteome analysis and reduce sample complexity, such as the study of sub-proteomes, namely exosomes, microparticles, released substances, organelles and membrane proteome [32] (Fig. 2). The pre-fractionation step, which will be performed before a comprehensive proteomic analysis, allows for the isolation of an organelle of interest or a specific chromatographic or electrophoretic fraction, facilitating the interpretation of samples [33].

4 Metabolomics and Lipidomics Approaches for Platelet-Rich Plasma Researches

4.1 Omics

Omics approaches can be thought of as a "modular approach" for any biological process consisting of a set of measures, in which the corresponding behavior and function are examined overall, apart from individually examining certain components of it [34] (Fig. 3). The field of omics has largely been driven by technological advances that enable cost-effective, high-throughput analysis of biological

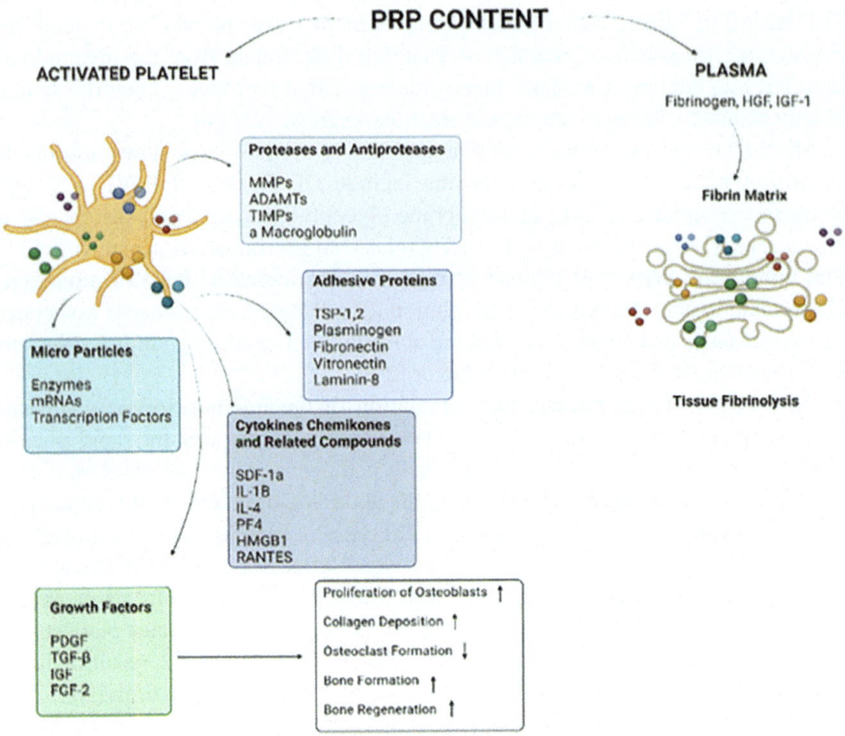

Fig. 2 PRP content, content and function of activated PRP, and plasma

molecules. It utilizes various technologies to aid in disease diagnosis and generate a comprehensive overview of human phenotypes and diseases [35].

Genomics, the first omics discipline to appear, focused on the study of whole genomes rather than "genetics," which interrogates individual variants or single genes

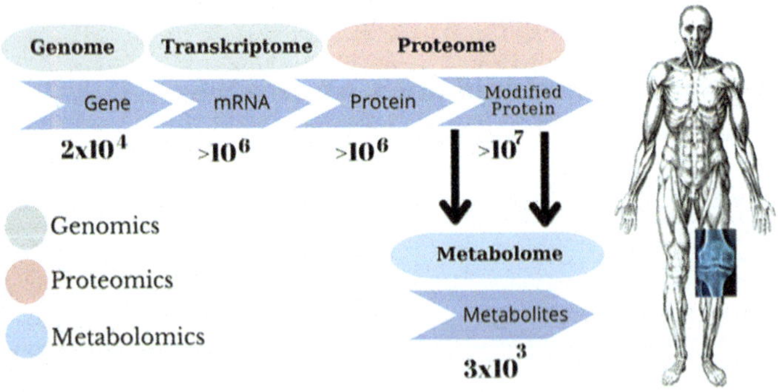

Fig. 3 Omics approaches (genomics, proteomics, metabolomics)

[36]. After genomic investigations, proteomic analysis involves the extensive examination of proteins, typically using biochemical methods. It involves characterizing the proteome, encompassing the expression, structure, functions, interactions, and modifications of proteins at any given stage. [36]. Today, the term proteomics covers most of the functional analysis of gene products or "functional genomics" [37]. Transcriptomic technologies are used to study an organism's transcriptome, which is the sum of all RNA transcripts. A transcriptome provides a current snapshot of all the transcripts found in a cell and offers genome-wide knowledge about gene structure and gene function to expose the molecular mechanisms involved in specific biological processes [38, 39].

4.2 Metabolomics

Metabolites are low molecular weight molecules such as amino acids, sugars, fatty acids, lipids and steroids that can be chemically converted during metabolism [40]. On the flip side, metabolomics is a subdivision of omics research focused on the high-throughput characterization of metabolites. It involves studying the metabolite composition of cell types, tissues, organs, or organisms. With advances in mass spectrometry, it has become possible to rapidly detect thousands of metabolites simultaneously even from only small amount of body fluids or tissue samples. Recent advances in device technology including the bioinformatics tools and software's enable comprehensive analysis of cellular metabolites (Fig. 4) [41]. This type of metabolomic analysis is increasingly being employed in various biological and medical research areas, spanning from bacteria to humans [42]. Minor changes in chemical structure and some external stimuli such as infection, trauma or pregnancy can significantly change the concentration of a metabolite in metabolic pathways. The reflection of pathological and traumatic effects on the phenotype can be determined by metabolomic analyzes [43] according to these changes.

4.3 Lipidomics

Lipids, which are typically hydrophobic biological substances, lack precise definition and, in most cases, soluble in organic solvents. Lipidomics is a subgroup of metabolomics. Analytical approaches used in lipidomics are comparatively different from the methods established for metabolomics. The majority of lipid molecules contain non-polar fatty acyl chains such as phospholipids and sphingolipids. These are specific physicochemical properties that should be considered during the development of assessment methods [44].

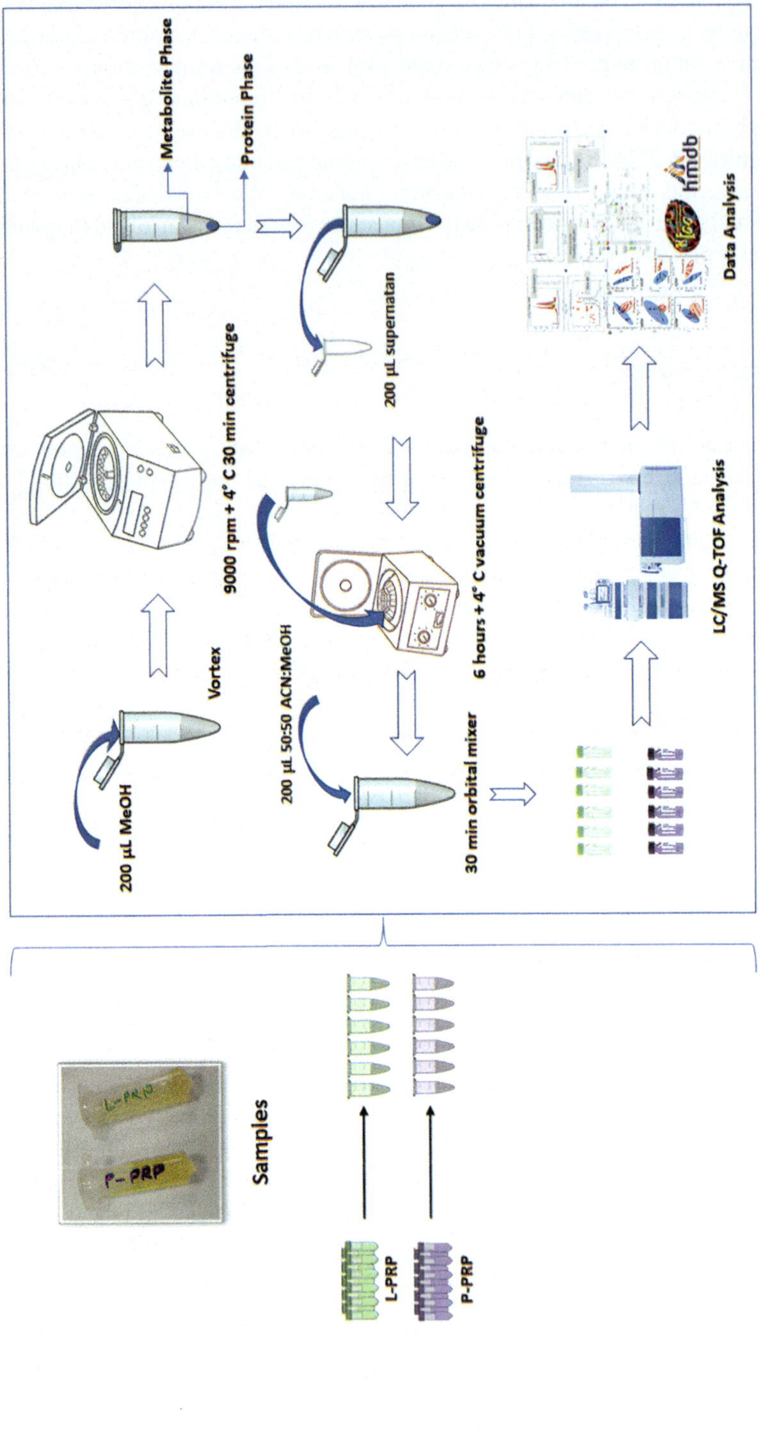

Fig. 4 Metabolomic analysis work flow

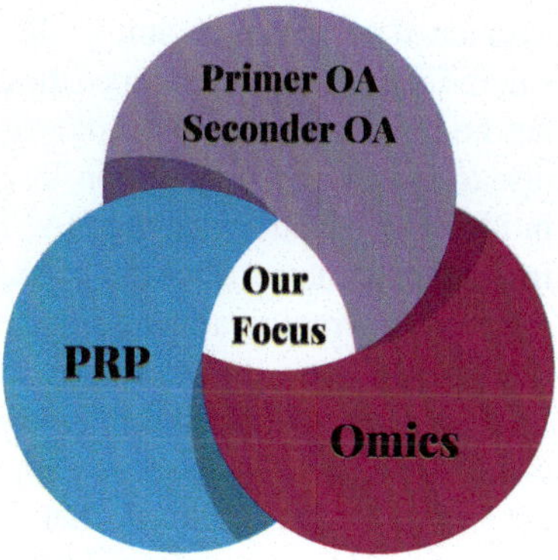

Fig. 5 Summary of the chapter focus

5 Literature Search and Article Selection

To address the urgent need for comprehensive research on OA and advanced biomolecular approaches, a thorough literature search was conducted using the WoS database on December 4, 2023. This search targeted topics including "Osteoarthritis, Platelet-rich plasma, proteomics, lipidomics, and metabolomic approaches," encompassing underlying terms (MesH terms) and keywords in the title and abstract (tiab). The scanning interval was set to the last five years (2019–2023) to ensure the inclusion of the most recent studies. Additionally, the electronic search was complemented by a manual selection of relevant journals and the reference lists of individual articles.

Studies were included if the article covered:

(1) Inclusion of omic studies on patients with OA.
(2) Articles examining PRP content with omic approaches.
(3) PRP methods applied to patients with KOA
(4) Articles with potential biomarker study in the differential diagnosis of OA.

Exclusion criteria included:

(1) Abstracts only and reviews.
(2) Lack of a non-OA control group.
(3) Non-English articles.

Articles examining the relationships between PRP, OA and omic approaches received a priority (Fig. 5). Focus was on the comparison of the molecular mechanisms of PRP and OA with studies including omic analysis.

6 Overview of Chapters

A total of 645 publications were reached when the publications in the literature were examined. Four hundred seventy-two of these publications were associated with OA and PRP. Nineteen were associated with PRP and omic studies. Hundred fifty were associated with OA and omic studies. Only four of them combined OA, PRP and omic studies. This finding developed the core of our rationale for this chapter. OA currently has no reliable and valid diagnostic biomarker that may detect the early onset of the disease and rapid progression in clinical practice. The clinical challenge persists in identifying and managing OA. Typically, achieved by physical examination and imaging techniques after the appearance of symptoms such as pain, deformity and stiffness. Current treatment modalities, whether they are oral medicines or intraarticular injections, aim to reduce these symptoms and preserve the joint function. These are mostly rescored by semi-quantitative questionnaires. Mapping metabolic changes is a comparatively new approach. The change in OA diagnosis and treatment paradigms can be improved when such new omic data derived from the literature can be integrated to our analyses [45].

6.1 Osteoarthritis-Platelet-Rich Plasma Omics

In a study conducted in 2021, the research aimed to assess the potential benefits of injecting 5% dextrose water into genicular nerves along with PRP for treating moderate KOA in elderly patients. Changes in synovial fluid protein concentrations were observed when 2D-PAGE electrophoresis was assessed. Brief results showed decreased inflammation and improved knee function for six months in the nerve block group. This suggested that combining dextrose water and PRP could offer a viable non-operative treatment for elderly KOA patients [46].

In a study [47] on the effectiveness of autologous platelet-rich plasma (aPRP) injections for mild OA but debating their impact on moderate to severe cases, the efficacy of PRP in grade two or higher OA was evaluated using proteomic analysis and the Lequesne index. Precise PRP injections guided by musculoskeletal ultrasound were administered into the intra-articular and simultaneous intra-articular and pes anserinus groups. Both groups exhibited decreased inflammation-related proteins. The second group showed superior results indicating the potential of PRP in treating moderate to severe KOA [47].

In a study [48] on extracellular vesicles (EVs), which are potential regulators of tissue repair, a significant increase in circulating platelet-derived EVs was observed after post submaximal strength rehabilitation in the examination of athletes recovering from muscle injuries. Label-free proteomics unveiled that 32% of the identified proteins were associated with defense and immunity, playing crucial roles in vesicle transport and wound healing through the isolation of EVs from PRP. These findings suggest a substantial regenerative impact of platelet-derived EVs in PRP, highlighting their therapeutic potential in tissue regeneration [48].

In another study [49] involving 19 elderly patients with moderate KOA and suprapatellar bursitis, PRP injections were not feasible due to medication constraints or surgical ineligibility. Intra-articular low-level laser therapy was administered every two weeks for three sessions. Ultrasound-guided low-level laser therapy resulted in a significant reduction in the Lequesne index, indicating improved knee function, sustained for at least three months. Suprapatellar synovial fluid volume and inflammation-associated proteins also significantly decreased, suggesting low-level laser therapy as a viable option for elderly KOA patient's ineligible for surgery or PRP injections [49]. These findings strengthen our hypothesis that a single diagnostic and treatment modality for OA cannot be defined with our current knowledge and sophisticated omic analysis may aid to identify better solutions.

6.2 *Osteoarthritis-Related Metabolomics*

In a study [50] among 82 metabolites examined, 42 OA-related metabolites were identified. Cystine, uric acid and tyrosine were identified as common metabolites associated with OA severity by performing advanced PLS-ROG (Partial least squares with rank order of groups) analysis. Large sample metabolome analyzes revealed that cysteine, an amino acid associated with antioxidant activity and glutamate homeostasis, may be a potential metabolic biomarker for radiographic OA diagnosis and early-phase synovitis.

In a study [51] of OA and rheumatoid arthritis (RA) patients, it was discovered that L-Pipecholic acid, deoxyguanosine, Lysophosphatidylcholine (LPC (16:0)) and oleamide may be a potential biomarker for oxidative stress and inflammation in end-stage OA and/or RA. Testosterone, androsterone sulfates, bile acids, histamine, L-arginine and oleamide have been presented as metabolites that may be useful targets for therapeutic manipulation.

In a study, as the first untargeted metabolomics study discussing the balance between obesity and KOA, was conducted among twenty-eight KOA patients (14 obesity (OKOA), 14 non-obesity (NOKOA) patients) and 15 healthy control groups [52]. For the circulation of metabolites, comparative analysis was attempted between the patient group and the control group and between OKOA and NOKOA. Twenty-one different metabolites are listed for the patient and control groups. These metabolites include PA(18:2(9Z,12Z)), PA(16:0/16:0), Phosphatidylethanolamine, Propane-1,3-diol, Sulfate, Benzoic acid, Phosphoric acid, Butyric acid, Acetic acid, There are L-Valine, L-Alanine, Stearic acid, Benzeneethanamine, Carbamic acid, Hydroxylamine, Indoleacetic acid, Urea, Glycine, Glycerol, Oleic acid and Lyso PC (18:2(9Z,12Z)). Listed are 15 significantly different metabolites recorded between OKOA and NOKOA. These metabolites include Phosphatidylethanolamine, PA(16:0/16:0), Lyso PC (18:2(9Z,12Z)), Stearic acid, Oleic Acid, Propane-1,3-diol, Carbamic acid, Hydroxylamine, L- Valine, Leucine, Isoleucine, L-Ornithine, D-Alanine, Butyric acid and Urea. 12 different metabolites were listed between the patient and control groups and between OKOA and NOKOA. These metabolites include Phosphatidylethanolamine, PA(16:0/16:0), Lyso PC (18:2(9Z,12Z)), Stearic acid, Oleic Acid, Propane-1,3-diol, Carbamic acid, Hydroxylamine, L- is available. There are Valine, Leucine, Butyric acid and Urea. With these 12 metabolites listed, the presence of acidic metabolites is particularly emphasized. Due to changes in these metabolites, 32 different pathways were affected for the patient and control groups, and 16 different pathways were affected for the OKOA and NOKOA groups. As a result of two comparison analyses: glycerolipid, nitrogen, glycerophospholipid, glycine, serine and threonine metabolism, tryptophan metabolism and fatty acids biosynthesis pathways are controlled as the most affected pathways. When the control group and the patient group are compared, phosphatidylethanolamine, lyso PC (18:2 (9Z,12Z)) and PA (16:0/16:0) levels are higher than the control. Comparison of the levels of these metabolites: The ratio between OKOA and NOKOA is that the levels are higher in OKOA. Carbamic acid and hydroxylamine levels were observed to be higher in OKOA patients, indicating that oxidative stress due to obesity is quite high. In KOA, a significant increase was observed in the level of propane-1,3-diol, which plays an important role in glycerolipid metabolism. It was also stated that glycerol and FFA were increased in the serum of KOA patients. It is studied in patient groups where glycine is lower and indoleacetic acids, the breakdown product of tryptophan, are much higher. The most critical point of this structure is the expansion of the governor. It has been suggested that valine is the only BCAA molecule with a significant difference and thus may be a biomarker for the diagnosis of KOA.

In a study conducted to determine the endotypes of OA. The study consists of a hip-knee prosthesis patient group consisting of 615 people and a control group consisting of 237 people [53]. Although patients with OA are older than the control group, there is no difference in gender distribution between the two groups. Metabolomic profiling was performed on fasting plasma samples with the kit containing the concentrations of 186 metabolites. Of the 186 metabolite concentrations measured, 162 were subjected to quality control criteria and were tested and included in the analysis. It has been stated that metabolomics data can be factorized using Bartlett's sphericity test that metabolomic data is sufficient for factor analysis with the Kaiser–Meyer–Olkin test, and that OA patients can be clustered with the Hopkins statistics. Accordingly, 615 OA patients were divided into 3 groups: A, B and C, according to their metabolomic factor scores. As a result of this separation, cluster A includes 66 patients, cluster B 200 and cluster C 349 patients with OA. The differences between these 3 clusters are based on butyrylcarnitine (C4), arginine and lysophosphatidylcholine, and palmitic acid. When patients in cluster A are compared with patients in other clusters and the control group, butyrylcarnitine (C4) concentration is much higher in patients in cluster A. High levels of butyrylcarnitine (C4) concentration are associated with muscle weakness. When patients in cluster B are compared with patients in other clusters and the control group, the arginine concentration is lower in patients in cluster B. When patients in cluster C are compared with patients in other clusters and the control group, the concentration of lysophosphatidylcholine (with palmitic acid) in cluster C is lower. Thus, these three clusters were associated with muscle weakness, arginine deficiency, and low inflammation, respectively. This study reports that there are three endotypes characterized by C4, arginine and lysophosphatidylcholine levels in plasma, thus developing a new approach to dividing OA patients into subtypes.

In this study, plasma samples of 12 OA patients and twenty healthy controls (HC) were collected and untargeted metabolomics detection from plasma was performed by mass spectrometry (GC/Q-TOF–MS) and gas chromatography [54]. The differential metabolites (DMs) between OA and HC were evaluated with multivariate analysis and a biomarker study that could be recommended for the diagnosis of OA was conducted. To understand the pathological infrastructure of OA, the synovial tissue transcriptome dataset containing 10 OA patients and 10 HCs was downloaded and differentially expressed genes (DEGs) were investigated. As the last step of the study, a joint analysis of DMs and DEGs was performed to investigate possible molecular mechanisms of OA. In the study, 58 metabolites were found to be significantly changed when OA patients and healthy individuals were compared. It has been suggested that especially succinic acid, xanthurenic acid and L-tryptophan may be helpful biomarkers for the diagnosis of OA.

6.3 Osteoarthritis-Related Proteomics

Yang et al. [55] Examined the correlation between menopause and knee KOA and investigated whether menopause influences the clinical recovery of the knee joint post OA and injury. Additionally, they aimed to examine whether menopause influences the clinical healing of the knee joint and to identify the mechanisms involved by analyzing the proteomic profiles of synovial tissue (ST) samples. Age and menopause were positively correlated with the protein profile of ST in female patients with OA or knee injury, suggesting that protein components may be affected by menopause. Postoperative clinical outcomes have been affected by menopause. It was concluded that menopause may regulate knee joint function, in part, by altering ST protein expression.

Acupotomy intervention, which is widely used in China, was used in rabbits with KOA [56]. The proteomic analyzes accumulated 68 different proteins associated with energy metabolism and Tricarboxylic acid (TCA) cycling. AIFM1 has been proposed to be an important regulator in the control of mitochondrial energy production. In addition, it has been suggested that FN1, VIM, COL12A1, COL14A1, MYBPH and DPYSL3 play an important role in acupotomy in the treatment of OA. In another proteomic study on rats, it was revealed that serpinA5 could be a potential biomarker for OA [57]. Only three studies were reached when the web site (https://clinicaltrials.gov/) of National Library of Medicine-Clinical Trials was accessed on December 14, 2023, with the key words of OA for the "condition/disease" filter and omics for the "other terms" section. A study was on acupuncture, which is not a routine method of OA treatment. The number of limited literature search results and clinical trials indicate that there is a need for more studies in OA omics. Focus on disease modifying approaches and cellular therapies could be a good approach to understand the changes after applications.

Dechêne et al. have investigated the potential of muscle-derived stem cells (mdMSCs) for treating OA using an equine model. Co-culturing mdMSCs with cartilage exposed to inflammatory cytokines showed a protective effect, reducing glycosaminoglycan release. Analysis revealed changes in decorin and matrix metalloproteinase 3 levels, indicating that mdMSCs may mitigate cartilage damage by modulating these factors, suggesting their potential for OA cell therapy [58].

6.4 Platelet-Rich Plasma-Related Omics

A recent proteomic study [59] on PRP that compared low leukocyte PRP (L-PRP), platelet poor plasma (PPP), and blood plasma revealed 50% of protein overlap. That study also identified 100 + pathways that were predominantly associated with inflammation. Major pathways, including acute phase response and coagulation, exhibited consistent regulation direction across experiments. This highlighted the need for standardized PRP formulations and emphasizes the role of inflammation-related pathways in autologous blood product mechanisms [59]. There are various PRP formulations and they do not present the same omic profiles.

The majority of studies in the literature primarily focus on comparing a single type of PRP used in the treatment of OA with alternative therapeutic approaches. However, there is a lack of metabolomic studies comparing different types of PRP's among themselves. Proteomic studies, on the other hand, are quite limited. Building upon this gap, a study has been designed to compare the metabolomic profiles of different types of fresh and frozen PRP. In this chapter, the methods and data of this study are presented.

6.5 Our Design and Pilot Study Outcomes

6.5.1 Preparation and Characterization of PRP

The pilot study was conducted with 6 male volunteers (age: 22 ± 1.7) after ethical board approval (Date: 05.04.2022-Decision number: 2022/14–57). 132 mL of blood was collected from each participant. The demographic information of the participants is presented in Table 1. Collected blood was separated into Leukocyte-rich platelet (L-PRP), Leukocyte-poor platelet (P-PRP).

For the preparation of P-PRP and L-PRP with a PRP separation device, 40 mL of blood was drawn from the antecubital vein into a syringe and was transferred to the citrate tube. This was centrifuged at 1200 g for 10 min at 22 °C (GLO GT416, Korea). After centrifugation, using the screw-type separation device of the device, the tube was turned upside down and sequentially PRP's were separated from the top of the tube with the help of the syringe. Approximately 10 mL portion at the top was the PPP. Approximately 3 mL portion below it was the P-PRP and the buffy coat layer, which was approximately 4 mL that remained above the erythrocyte layer was the L-PRP (Fig. 6).

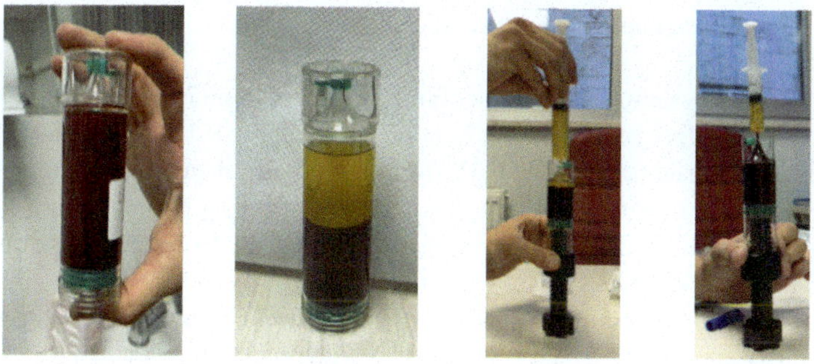

a b c d

Fig. 6 Separation of plasma, P-PRP, and L-PRP using a PRP separation kit. **a** blood collection in a citrate tube, **b** after centrifugation, **c** P-PRP, **d** L-PRP

Table 1 Demographic characteristics of the participants were summarized as mean ± standard deviation and percentage values

Variables*		n	Mean ± SD
Gender		6	Male
Age		6	22 ± 1.7
Height		6	178.8 ± 6.2
Weight		6	81.17 ± 13.2
BMI		6	26 ± 5%
Additional disease	Yes	1	% 16.6
	No	5	% 83.3
Smoking user	User	1	% 16.6
	Non-user	5	% 83.3

*Variables were presented as mean ± standard deviation (SD) and percentage values
BMI: Body Mass Index

6.5.2 Extraction of Metabolites from PRP Samples

From each prepared PRP sample, (P-PRP, L-PRP), 0.1 mL was taken into a microcentrifuge tube (Eppendorf, Germany). To this, 0.2 mL of methanol:water (70:30, v/v) was added. After vortexing for a minute, it was centrifuged at 10,000 rpm for 10 min at + 4 °C in a chilled centrifuge (Hettich Universal 320 R, Germany). After centrifugation, 0.1 mL was taken from the upper phase into a new microcentrifuge tube. The collected samples were evaporated in a vacuum centrifuge (Labconco, CentiVap, 7,310,030, US) until the solvents completely evaporated.

6.5.3 Chromatographic Conditions

Q-TOF LC/MS system, Analytical column: C18 Column (2.5 μm particle size, 2.1 × 100 mm, Xbridge) Mobile phase A: water (containing 0.1% formic acid), Mobile phase B: acetonitrile (containing 0.1% formic acid), gradient elution; time program [(A%: 90 (1 min) → 65% (4 min) → 10% (11 min) → 90% (14 min) → 90% (20 min)]. Flow rate: 0.2 mL/min., Injection volume: 5 μL, Column oven temperature: 35 °C. MS conditions: Electrospray ionization (ESI-Positive mode).

Injection plan: After column conditioning for chromatographic analysis, prepared fresh and frozen samples, H, and QC samples were injected separately. Mobile phase and QC samples were injected every 10 samples to ensure control and calibration.

6.5.4 Statistical Data Analysis

The raw data obtained from chromatographic analysis were converted into mzML file format. Then, for each injection performed using the MZmine (MZmine-2.53) software program, mass-to-charge (m/z) ratios, retention times, and peak areas were determined. The identified values were converted to an Excel table for statistical comparisons. Before proceeding to statistical data comparison, a normalization process was performed. Normalization was undertaken by dividing the peak area of each injection by the average peak areas of that peak. After the normalization process, the averages of repeated injections for each fresh and frozen sample were calculated. From the average results, t-tests, fold change (FC) values were calculated. A filtering process was applied to determine how much fresh samples differed from frozen samples and in which direction the change occurred (increased, decreased). The applied filters were blank ≥ 3 (result of dividing the average peak areas of fresh samples by the average peak areas of the mobile phase), t-test < 0.05, and FC ≥ 2. The number of peaks obtained as a result of each filtering step was recorded (Table 2).

The m/z values corresponding to the t-test and FC values obtained from the filtering process were saved in a new Excel file. The saved file was loaded into the MetaboAnalyst (MetaboAnalyst 5.0) software program to identify metabolites that vary between groups. To detect changes between fresh and frozen groups, data containing normalized peak areas for all samples and the m/z values for these peaks were saved in a new Excel file. The prepared Excel file was loaded into the

Table 2 Number of peaks (L-PRP/P-PRP): The total number of peaks obtained after processing the raw data, followed by the number of remaining peaks after sequential application of the t-test

	Number of peaks detected in our experimental conditions (L-PRP/P-PRP)
Total peak	8817
t-test < 0.05	3059
FC ≥ 2	1692

MetaboAnalyst (MetaboAnalyst 5.0) software program and converted into a Principal Component Analysis (PCA) graph.

6.5.5 Findings

The normalized data were processed using MetaboAnalyst 5.0. and PCA was performed (Fig. 7).

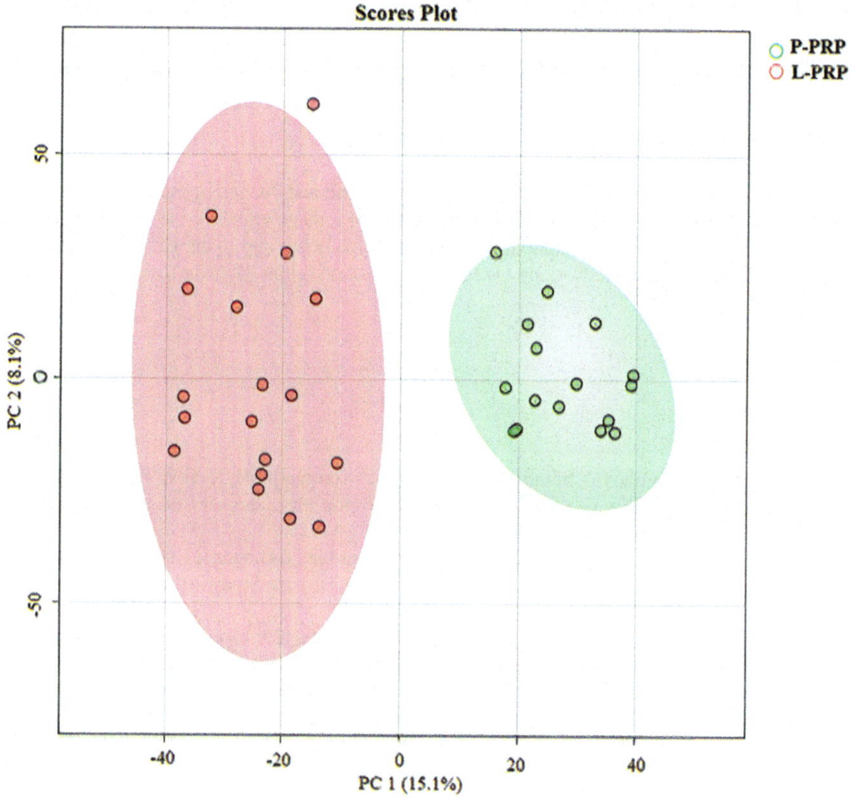

Fig. 7 Principal Component Analysis of PRP types

7 Concluding Remarks

This comprehensive study provides valuable insights into the intricate relationship between PRP, OA, and various omic approaches, including proteomics, metabolomics, and lipidomics. The research delves into the complexities of KOA, a prevalent and debilitating chronic disorder affecting millions globally.

In conclusion, the study contributes significantly to the understanding of KOA, its association with sport-related injuries, and the therapeutic potential of PRP. The utilization of omics approaches enhances our comprehension of the molecular mechanisms underlying the disease and PRP intervention. The findings also underscore the importance of fresh PRP for optimal physiological impact. This research lays the groundwork for future studies aiming to refine PRP formulations and advance our knowledge of OA at the molecular level, potentially paving the way for more effective diagnostic and therapeutic strategies.

Acknowledgment This study was funded by the Scientific and Technological Research Council (TÜBİTAK) Project #222S093. Feza Korkusuz MD is an active member of the Turkish Academy of Sciences (TÜBA). Presented outcomes on osteoarthritis is relevant to the EU-COST-CA21110 NetwOArk-The European Network on OsteoArthritis Action (https://netwoark.eu/).

References

1. Constantino de Campos G, Mundi R, Whittington C, Toutounji MJ, Ngai W, Sheehan B (2020) Osteoarthritis, mobility-related comorbidities and mortality: an overview of meta-analyses. Therapeutic Adv Musculoskele Dis 12:1759720X20981219
2. GBD (2019) Global burden of 369 diseases and injuries in 204 countries and territories, 1990–2019: a systematic analysis for the Global Burden of Disease Study 2019. https://vizhub.healthdata.org/gbd-results/
3. Quicke J, Conaghan P, Corp N, Peat G (2022) Osteoarthritis year in review 2021: epidemiology and therapy. Osteoarthritis Cartilage 30(2):196–206
4. Cui A, Li H, Wang D, Zhong J, Chen Y, Lu H (2020) Global, regional prevalence, incidence and risk factors of knee osteoarthritis in population-based studies. EClinicalMedicine 29
5. Yahaya I, Wright T, Babatunde OO, Corp N, Helliwell T, Dikomitis L, Mallen CD (2021) Prevalence of osteoarthritis in lower middle-and low-income countries: a systematic review and meta-analysis. Rheumatol Int 41(7):1221–1231
6. Organization Wh (2022) Musculoskeletal health. https://www.who.int/news-room/fact-sheets/detail/musculoskeletal-conditions
7. Precention CFDCA (2020) Osteoarthritis (OA). https://www.cdc.gov/arthritis/basics/osteoarthritis.htm
8. Roos EM, Arden NK (2016) Strategies for the prevention of knee osteoarthritis. Nat Rev Rheumatol 12(2):92–101
9. Shamekh A, Alizadeh M, Nejadghaderi SA, Sullman MJ, Kaufman JS, Collins GS, Kolahi A-A, Safiri S (2022) The burden of osteoarthritis in the Middle East and North Africa region from 1990 to 2019. Front Med 9:881391
10. Kingsbury SR, Gross HJ, Isherwood G, Conaghan PG (2014) Osteoarthritis in Europe: impact on health status, work productivity and use of pharmacotherapies in five European countries. Rheumatology 53(5):937–947

11. Bikbov MM, Kazakbaeva GM, Gilmanshin TR, Zainullin RM, Rakhimova EM, Fakhretdinova AA, Tuliakova AM, Rusakova IA, Panda-Jonas S, Nuriev IF (2022) Prevalence and associated factors of osteoarthritis in the Ural Eye and Medical Study and the Ural Very Old Study. Sci Rep 12(1):12607
12. Turkiewicz A, Petersson IF, Björk J, Hawker G, Dahlberg LE, Lohmander LS, Englund M (2014) Current and future impact of osteoarthritis on health care: a population-based study with projections to year 2032. Osteoarthritis Cartilage 22(11):1826–1832
13. Ratneswaran A, Kapoor M (2021) Osteoarthritis year in review: genetics, genomics, epigenetics. Osteoarthritis Cartilage 29(2):151–160
14. Heidari B (2011) Knee osteoarthritis prevalence, risk factors, pathogenesis and features: Part I. Caspian J Intern Med 2(2):205
15. Hertel G, Hochrein A, Suren C, Minzlaff P, Banke I, Willers J, von Eisenhart-Rothe R, Prodinger P (2022) Injury incidence and specific injury patterns in app-based bodyweight training (Freeletics): results of an international survey with 3668 participants. BMC Sports Sci Med Rehabil 14(1):1–8
16. Prien A, Boudabous S, Junge A, Verhagen E, Delattre BM, Tscholl PM (2020) Every second retired elite female football player has MRI evidence of knee osteoarthritis before age 50 years: a cross-sectional study of clinical and MRI outcomes. Knee Surg Sports Traumatol Arthrosc 28:353–362
17. Palmieri-Smith RM, Cameron KL, DiStefano LJ, Driban JB, Pietrosimone B, Thomas AC, Tourville TW, Consortium ATO (2017) The role of athletic trainers in preventing and managing posttraumatic osteoarthritis in physically active populations: a consensus statement of the Athletic Trainers' Osteoarthritis Consortium. J Athletic Train 52(6):610–623
18. Arant KR, Katz JN (2022) Epidemiology of post-traumatic osteoarthritis of the lower extremity: premature aging of youthful joints. In: Early osteoarthritis: state-of-the-art approaches to diagnosis, treatment and controversies, pp 39–49
19. Drawer S, Fuller C (2001) Propensity for osteoarthritis and lower limb joint pain in retired professional soccer players. Br J Sports Med 35(6):402–408
20. Lohmander LS, Englund PM, Dahl LL, Roos EM (2007) The long-term consequence of anterior cruciate ligament and meniscus injuries: osteoarthritis. Am J Sports Med 35(10):1756–1769
21. Krishnasamy P, Hall M, Robbins SR (2018) The role of skeletal muscle in the pathophysiology and management of knee osteoarthritis. Rheumatology 57(suppl_4):iv22–iv33
22. Jarraya M, Roemer F, Kwoh CK, Guermazi A (2023) Crystal arthropathies and osteoarthritis—where is the link? Skeletal Radiol 52(11):2037–2043
23. Kaeley GS, Bakewell C, Deodhar A (2020) The importance of ultrasound in identifying and differentiating patients with early inflammatory arthritis: a narrative review. Arthritis Res Ther 22(1):1–10
24. Collins T, Alexander D, Barkatali B (2021) Platelet-rich plasma: a narrative review. EFORT Open Rev 6(4):225–235
25. Çiftçi E, Demirdis I, Korkusuz F (2023) Regenerative and restorative treatments of the musculoskeletal system: focus on platelet-rich plasma (PRP) in osteoarthritis. In: Beauty, aging, and antiaging. Elsevier, pp 413–427
26. Wasterlain AS, Braun HJ, Dragoo JL (2016) Contents and formulations of platelet rich plasma. In: Platelet Rich Plasma in musculoskeletal practice, pp 1–29
27. Yun SH, Sim EH, Goh RY, Park JI, Han JY (2016) Platelet activation: the mechanisms and potential biomarkers. BioMed Res Int
28. Shevchuk O, Begonja AJ, Gambaryan S, Totzeck M, Rassaf T, Huber TB, Greinacher A, Renne T, Sickmann A (2021) Proteomics: a tool to study platelet function. Int J Mol Sci 22(9):4776
29. Lee HW, Choi K-H, Kim J-Y, Kim K-O, Haotian B, Yuxuan L, Noh K-C (2020) Proteomic classification and identification of proteins related to tissue healing of platelet-rich plasma. Clin Orthop Surg 12(1):120–129
30. Sung P-S, Huang T-F, Hsieh S-L (2019) Extracellular vesicles from CLEC2-activated platelets enhance dengue virus-induced lethality via CLEC5A/TLR2. Nat Commun 10(1):2402

31. Dalrymple N, Mackow ER (2011) Productive dengue virus infection of human endothelial cells is directed by heparan sulfate-containing proteoglycan receptors. J Virol 85(18):9478–9485
32. Vélez P, García Á (2015) Platelet proteomics in cardiovascular diseases. Transl Proteom 7:15–29
33. Zufferey A, Fontana P, Reny JL, Nolli S, Sanchez JC (2012) Platelet proteomics. Mass Spectrom Rev 31(2):331–351
34. Ge H, Walhout AJ, Vidal M (2003) Integrating 'omic' information: a bridge between genomics and systems biology. Trends Genet 19(10):551–560
35. Karczewski KJ, Snyder MP (2018) Integrative omics for health and disease. Nat Rev Genet 19(5):299–310
36. Hasin Y, Seldin M, Lusis A (2017) Multi-omics approaches to disease. Genome Biol 18(1):1–15
37. Pandey A, Mann M (2000) Proteomics to study genes and genomes. Nature 405(6788):837–846
38. Dong Z, Chen Y (2013) Transcriptomics: advances and approaches. Sci China Life Sci 56:960–967
39. Lowe R, Shirley N, Bleackley M, Dolan S, Shafee T (2017) Transcriptomics technologies. PLoS Comput Biol 13(5):e1005457
40. Patti GJ, Yanes O, Siuzdak G (2012) Metabolomics: the apogee of the omics trilogy. Nat Rev Mol Cell Biol 13(4):263–269
41. Chen Y, Li E-M, Xu L-Y (2022) Guide to metabolomics analysis: a bioinformatics workflow. Metabolites 12(4):357
42. Mechanick JI, Adams S, Davidson JA, Fergus IV, Galindo RJ, McKinney KH, Petak SM, Sadhu AR, Samson SL, Vedanthan R (2019) Transcultural diabetes care in the United States–a position statement by the American Association of Clinical Endocrinologists. Endocr Pract 25(7):729–765
43. Scrivo R, Casadei L, Valerio M, Priori R, Valesini G, Manetti C (2014) Metabolomics approach in allergic and rheumatic diseases. Curr Allergy Asthma Rep 14:1–11
44. Holčapek M, Liebisch G, Ekroos K (2018) Lipidomic analysis. Publications
45. Pertusa C, Mifsut D, Morales JM, Tarín JJ, Cano A, Monleón D, García-Pérez MÁ (2022) Metabolomic analysis of severe osteoarthritis in a Spanish population of women compared to healthy and osteoporotic subjects. Metabolites 12(8):677
46. Chen JL, Chen CH, Cheng CH, Chen CC, Lin KY, Chen CP (2021) Can the addition of ultrasound-guided genicular nerve block using 5% dextrose water augment the effect of autologous platelet rich plasma in treating elderly patients with knee osteoarthritis? Biomed J 44(6):S144–S153
47. Chen CP, Chen J-L, Hsu C-C, Pei Y-C, Chang W-H, Lu H-C (2019) Injecting autologous platelet rich plasma solely into the knee joint is not adequate in treating geriatric patients with moderate to severe knee osteoarthritis. Exp Gerontol 119:1–6
48. Catitti G, Cufaro MC, De Bellis D, Cicalini I, Vespa S, Tonelli F, Miscia G, Secondi L, Simeone P, De Laurenzi V (2022) Extracellular vesicles in regenerative processes associated with muscle injury recovery of professional athletes undergoing sub maximal strength rehabilitation. Int J Mol Sci 23(23):14913
49. Chen JL, Hsu CC, Chen WC, Peng YN, Chen CP, Suputtitada A (2022) Intra-articular laser therapy may be a feasible option in treating knee osteoarthritis in elderly patients. BioMed Res Int
50. Sasaki E, Yamamoto H, Asari T, Matsuta R, Ota S, Kimura Y, Sasaki S, Ishibashi K, Yamamoto Y, Kami K (2022) Metabolomics with severity of radiographic knee osteoarthritis and early phase synovitis in middle-aged women from the Iwaki Health Promotion Project: a cross-sectional study. Arthritis Res Ther 24(1):1–11
51. Nieminen P, Hämäläinen W, Savinainen J, Lehtonen M, Lehtiniemi S, Rinta-Paavola J, Lehenkari P, Kääriäinen T, Joukainen A, Kröger H (2022) Metabolomics of synovial fluid and infrapatellar fat pad in patients with osteoarthritis or rheumatoid arthritis. Inflammation 45(3):1101–1117
52. Senol O, Gundogdu G, Gundogdu K, Miloglu FD (2019) Investigation of the relationships between knee osteoarthritis and obesity via untargeted metabolomics analysis. Clin Rheumatol 38:1351–1360

53. Werdyani S, Liu M, Zhang H, Sun G, Furey A, Randell EW, Rahman P, Zhai G (2021) Endotypes of primary osteoarthritis identified by plasma metabolomics analysis. Rheumatology 60(6):2735–2744
54. Huang Z, He Z, Kong Y, Liu Z, Gong L (2020) Insight into osteoarthritis through integrative analysis of metabolomics and transcriptomics. Clin Chim Acta 510:323–329
55. Yang X, Fu Z, Ren Q, Wang S, Yin S, Liu J (2022) Correlation of synovial tissue protein abundance with menopause in osteoarthritis. Am J Transl Res 14(10):7403
56. Liu J, Zeng W, Lin Q, Dai R, Lu L, Guo Z, Lian X, Pan X, Liu H, Xiu ZB (2022) Proteomic analyses reveals the mechanism of acupotomy intervention on the treatment of knee osteoarthritis in rabbits. In: Evidence-based complementary and alternative medicine
57. Zhu T, Tan Q, Xin X, Li F, Zhang K, Liu Z, Tian Y (2022) Proteomic analysis of human articular cartilage unravels the dyscoagulation in osteoarthritis and the potential value of serpinA5 as a biomarker for osteoarthritis. PROTEOMICS–Clinic Appl 16(3):2100117
58. Dechêne L, Colin M, Demazy C, Fransolet M, Niesten A, Arnould T, Serteyn D, Dieu M, Renard P (2023) Characterization of the proteins secreted by equine muscle-derived mesenchymal stem cells exposed to cartilage explants in osteoarthritis model. Stem Cell Rev Rep 19(2):550–567
59. Miroshnychenko O, Chalkley RJ, Leib RD, Everts PA, Dragoo JL (2020) Proteomic analysis of platelet-rich and platelet-poor plasma. Regenerative therapy 15:226–235

Bilge Başak Fidan is a master's student in the Department of Bioengineering at Hacettepe University, Turkey. She holds a dual undergraduate degree in Biology Education from Gazi University and Management Information Systems from Anadolu University. As part of the Erasmus program, she studied for one year in the Department of Biotechnology at the University of Cádiz, Spain. Her research focuses on **Q-TOF LC MS analysis, musculoskeletal system diseases, proteomics, and metabolomics.** She has worked extensively with biological samples, including **blood plasma, urine, cell culture, and tissue,** as well as pharmaceutical formulations such as **tablets, nanoformulations, and parenteral solutions.** Additionally, she has experience with **micro-LC techniques.** She is currently an **Assistant Specialist at the Project Management and Coordination Office of the Scientific and Technological Research Council of Turkey (TÜBİTAK).**

Murat Koçak (https://www.linkedin.com/in/muratkocak/) is a specialist in Medical Informatics and Bibliometrics with extensive experience in academic data management and health informatics. He completed his PhD in Medical Informatics at Middle East Technical University (METU) in 2023, following his MSc in the same field. During his doctoral studies, he conducted research at the University of Carlos III Madrid, Spain (2018–2019) with TUBITAK 2214-A programme scholarship. He holds a BSc in Healthcare Management from Baskent University. Since 2009, Dr. Koçak has been serving as the Web Coordinator for the University Ranking by Academic Performance (URAP) project. He currently works as a Lecturer at Baskent University's Faculty of Medicine, Department of Medical Informatics. His research interests span bibliometrics analysis, health informatics, neuroscience, content management systems, and web development. Dr. Koçak has made significant contributions to various international projects, including the project for which he won the first prize at TEKNOFEST- Artificial Intelligence Competition in Health project on mammography image analysis using computer vision. Throughout his career, he has been actively involved in numerous e-learning projects and has contributed to multiple international conferences in medical informatics and health administration.

İlayda Demirdiş graduated from Başkent University, Department of Molecular Biology and Genetics in June 2021. She started her master's degree in Molecular Biology Department of Hacettepe University Biology Department in February 2023 and is still continuing her education. Her research interests include molecular biology, genetic and metabolomics.

Hakan Aydınlı graduated as the top student from Nesibe Aydın High School, Ankara and began his medical education at Hacettepe University Faculty of Medicine in September 2020. In July–August 2024, he completed a medical observership at the Mayo Clinic Florida, Department of General Surgery. Currently, he is a research intern in the Department of Sports Medicine, focusing on treatment modalities for osteoarthritis. He continues his education as a medical student while actively engaging in clinical and research activities.

Emine Çiftci MD attended medical medical specification in Sports Medicine at Hacettepe Medical Faculty, Department of Sports Medicine between 2020 and 2025, after graduating from Ankara Yıldırım Beyazıt University Medical Faculty in 2018. Between 2018 and 2020 she studied in Ambulance Emergency Doctor (911). Dr. Emine Çiftci Özotuk completed her medical specialization in Sports Medicine at Hacettepe University Medical Faculty, Department of Sports Medicine between 2020 and 2025, following her graduation from Ankara Yıldırım Beyazıt University Medical Faculty in 2018. Between 2018 and 2020, she worked as an Ambulance Emergency Doctor (911), where she gained extensive experience in pre-hospital emergency care, trauma management, and critical patient stabilization. Her clinical and research interests focus on sports injuries, rehabilitation, athlete mental health. Dr. Emine Çiftci Özotuk is particularly interested in regenerative therapies in sports medicine and the use of musculoskeletal ultrasound (MSK USG) for diagnosis and treatment guidance in sports-related injuries. Dr. Emine Çiftci Özotuk has served as a Tournament Doctor for Volleyball Nations League (VNL) Women's Competitions between 2022 and 2024 and as a National Team Doctor for football, rugby, and ice hockey, providing medical support in international tournaments, including the UEFA Women's European Championships, Rugby U18 Women's Trophybetween 2021 and 2025.

Ozan Kaplan PhD., is working as researcher in Faculty of Pharmacy, Department of Analytical Chemistry at Hacettepe University. His undergraduate from chemistry department, and PhD. from analytical chemistry. His main research areas are mass spectrometry based metabolomics, lipidomics and proteomics where he employ method development and bioinformatical approaches to address various clinical research questions.

Özge Boyacıoğlu PhD (https://scholar.google.com/citations?user=6UleqH4AAAAJ&hl=tr) completed her Master's and doctoral degrees in Bioengineering at Hacettepe University between 2017 and 2025, after graduating as the highest-ranked student of the Faculty of Science from Hacettepe University Faculty of Science Department of Biology between 2013 and 2017. She completed a summer internship at Karolinska Institute, Department of Physiology and Pharmacology, Division of Experimental Audiology in Sweden in 2016. She has been working as a research assistant at Atılım University Faculty of Medicine, Department of Medical Biochemistry since 2019. She received a bronze medal in the 4th Istanbul International Invention Fair 2019 (ISIF'19) Invention Awards by the Turkish Patent and Trademark Office in 2019 and the Best Scientific Research Award in Basic Medical Sciences among XIII. Dr. Aysun-Ahmet Küçükel Medicine Awards by Güven Hospital and Güven Education and Health Foundation in 2021. She has been taking part in research projects constituting cancer therapies, bioengineering and tissue engineering approaches and drug development. Her research interests focus on cell and cancer cell biology, biomaterial-cell interactions, drug development, nanoparticle delivery systems, targeted therapies, non-small cell lung cancer and cannabinoids. She is now an active member of the European Association for Cancer Research (EACR), the Molecular Cancer Research Association (MOKAD) and the European Society for Medical Oncology (ESMO).

Petek Korkusuz M.D. Ph.D., Professor of Histology and Embryology

Hacettepe University Faculty of Medicine, Department of Histology & Embryology, and METU MEMS Center, Ankara Turkey

Petek Korkusuz received her MD degree from Gazi University Faculty of Medicine in 1991. She completed her residency in Histology and Embryology at Hacettepe University Faculty of Medicine in 1997 and received her Ph.D. in 2000 from Gazi University Faculty of Medicine Department of Medical Pharmacology Her professional appointments include Assistant Professor position in 1998, Associate Professor (award of tenure) in 2003 and full Professor in 2010 (present position) in Hacettepe University Faculty of Medicine Department of Histology and Embryology. Her secondary appointments are in Department of Nanomedicine and Nanotechnology, Department of Bioengineering, Department of Stem Cell Sciences and, Department of Oral Biology and METU MEMS Center. She received the Turkish Scientific and Research Council (TUBİTAK-BAYG) National Doctoral Scholarship between 1991 and 1993. She has six patents, 3131 citations for 186 publications in Science Citation Index (WOS) with an h index of 31. She has 5515 citations in Google Scholar with an h index of 39 on January 2025.

Mustafa Çelebier is a leading academic in the fields of analytical chemistry and metabolomics in Turkey. Born on January 17, 1981, he began his academic journey by earning his bachelor's degree in Chemistry from the Faculty of Science at Ankara University in 2001. He then pursued his first master's degree at the Ankara University Institute of Natural Sciences Department of Chemistry before joining Hacettepe University's Faculty of Pharmacy, Department of Analytical Chemistry as a Research Assistant in 2004. Continuing his career at Hacettepe University, Mustafa Çelebier completed his second master's degree in 2007 and earned his PhD in 2013 from the Health Sciences Institute's Analytical Chemistry program. During the 2010–2011 period, he further enhanced his expertise by conducting thesis studies on metabolomics and proteomics at Madrid Autonomous University under the YÖK Doctoral Research Scholarship. His academic progression is marked by significant milestones. He began as an Assistant Professor in 2013, was promoted to Associate Professor in 2017, and achieved the title of Professor in 2023. His research, which primarily focuses on drug analysis and metabolomics, has gained recognition both nationally and internationally through numerous publications and project contributions.In addition to his research, Mustafa Çelebier is deeply involved in teaching, supervising graduate theses, organizing scientific meetings, and performing various institutional duties. At Hacettepe University, he has held management roles in centers such as HÜNİTEK and HÜNİKAL. Throughout his career, he has actively participated in national and international projects, often as a principal investigator or key researcher, with support from institutions such as TUBITAK and BAB. His work on developing advanced analytical methods—especially those involving LC-MS based untargeted metabolomic approaches and proteomics—has been widely recognized.His scientific contributions have been further acknowledged through awards, including:

- Hacettepe University Science and Incentive Award (2017)
- Turkish Pharmacists Association Pharmacy Academy Science and Incentive Award (2019)

Mustafa Çelebier continues to make contributions to the fields of analytical chemistry and metabolomics.

Feza Korkusuz MD (https://scholar.google.com/citations?user=JtEsNWQAAAAJ&hl=en)(https://avesis.hacettepe.edu.tr/feza.korkusuz) attended medical specification in Orthopedic Surgery and Traumatology at Gazi University Medical Faculty, Department of Orthopedic Surgery and Traumatology between 1987 and 1992, after graduating from Ankara University Medical Faculty in 1986. Between 1989 and 1990 he studied in Osaka University Faculty of Medicine, Department of Orthopaedic Surgery from where he graduated with a postgraduate diploma. His clinical and research studies focused on Spine Surgery, Sports Medicine and Basic Research. Dr. Korkusuz worked at the Middle East Technical University (METU) as an orthopedic surgeon and Medical Director at the Medical Center, and at the Department of Physical Education and Sports as the Head of the Department until 2013. In 1994 and in 2001 he became Associate Professor and Professor. In 1999 and 2000, he received the METU Prof. Dr. N. Parlar Foundation young investigator and the Scientific and Technological Research Council of Turkey (TÜBİTAK) Promotion Award in Medical Sciences. Between 1991 and 2016, he was a member to the Orthopaedic Research Society. In 2008 he became an active member of the Association of Bone and Joint Surgeons (ABJS), and since 2006 he has been the corresponding and deputy editor of Clinical Orthopaedics and Related Research (CORR). In 2017, he became an active member to the Turkish Academy of Science (TUBA). He was the Advisor to the President of TÜBİTAK between 2018 and 2021. He currently is the Chairman of the Department of Hacettepe University Medical Faculty, Department of Sports Medicine. His current research interests are motion analysis, multi-omics, novel biosensors for osteoarthritis diagnosis and new therapeutics delivery using micro-needles.

Synovial Joint Fluid Omics Researches in Osteoarthritis: An Affinity Chromatography-Based Perspective

Serhat Aladağ, Ilayda Demirdiş, Burcu Gökçal Kapucu, Bilge Başak Fidan, Batuhan Erhan Aktaş, Merve Gizer, Ozan Kaplan, Idil Yet, Melis Şardan Ekiz, Petek Korkusuz, Mustafa Çelebier, Ömür Çelikbıçak, Ali Tuncel, and Feza Korkusuz

Abstract Synovial joint fluid is the lubricant and nutrient of articular joint cartilage. Osteoarthritis research recently focused on genetics, epigenetics, proteomics, metabolomics and lipidomics. This technology not only allows the identification of

Supplementary Information The online version contains supplementary material available at https://doi.org/10.1007/978-981-96-3808-6_3.

S. Aladağ
Department of Bioengineering, Hacettepe University, Ankara, Türkiye
e-mail: serhataladag44@gmail.com

I. Demirdiş
Department of Biology, Hacettepe University, Ankara, Türkiye
e-mail: ilaydademirdis@gmail.com

B. G. Kapucu · A. Tuncel
Department of Chemical Engineering, Hacettepe University, Ankara, Türkiye
e-mail: burcugokcal@gmail.com

A. Tuncel
e-mail: atuncel0@gmail.com

B. B. Fidan · O. Kaplan · M. Çelebier
Department of Basic Pharmacy Sciences, Hacettepe University, Ankara, Türkiye
e-mail: bilgefidan98@gmail.com

O. Kaplan
e-mail: ozankaplan@hacettepe.edu.tr

M. Çelebier
e-mail: mcelebier@gmail.com

B. E. Aktaş · F. Korkusuz (✉)
Faculty of Medicine, Department of Sports Medicine, Hacettepe University, Ankara, Türkiye
e-mail: feza.korkusuz@hacettepe.edu.tr

B. E. Aktaş
e-mail: batuhanerhanaktas@gmail.com

N. Sağlam et al. (eds.), *Nano-Biomaterials in Tissue Repair and Regeneration*, Tissue Repair and Reconstruction, https://doi.org/10.1007/978-981-96-3808-6_3

phenotypes and endotypes but may also monitor the clinical, molecular and therapeutic stages of the disease. In this review, we summarized biomarker-based diagnostics of osteoarthritis. Our focus was the synovial joint fluid. Affinity chromatographic and omic-based strategies are presented. The limitations were highlighted. Insight into synovial joint fluid processing is provided. IMAC sorbent synthesis is discussed. LC–MS/MS capability after phosphopeptide enrichment is described. Outcomes of our pilot study are presented. The isolation and detection capabilities of the IMAC sorbent for other phosphostructures and their clinical use will be examined in future.

Keywords Osteoarthritis · Synovial joint fluid · Affinity column chromatography · Mass spectrometry · Proteomics · Metabolomics · Lipidomics · Omics · Biomarkers

1 Background Information: Osteoarthritis

Incidence of osteoarthritis (OA) has increased by 9% in the last 28 years [1]. Prevalence of OA is reported to be 16.1% and it is more common in low- and middle-income countries [2, 3]. It has been reported that 240 million individuals worldwide have symptomatic OA and this rate is especially higher in women over 60 years of age (Table 1). Knee joint OA was detected in 654.1 million people over the age of 40 in 2022 and its incidence was reported to be 203 per 10 thousand people [2]. The

M. Gizer
Department of Stem Cell Science, Hacettepe University, Ankara, Türkiye
e-mail: gizermerve@gmail.com

M. Gizer · P. Korkusuz
MEMS Center, Middle East Technical University, Ankara, Türkiye
e-mail: petekkorkusuz68@gmail.com

I. Yet
Department of Bioinformatics, Hacettepe University, Ankara, Türkiye
e-mail: idilyet@gmail.com

M. Ş. Ekiz · Ö. Çelikbıçak
Department of Chemistry, Hacettepe University, Ankara, Türkiye
e-mail: melisardan@yahoo.com

Ö. Çelikbıçak
e-mail: omurcelikbicak@gmail.com

P. Korkusuz
Faculty of Medicine, Department of Histology and Embryology, Hacettepe University, Ankara, Türkiye

Ö. Çelikbıçak
Advanced Technologies Application and Research Center, Hacettepe University, Ankara, Türkiye

prevalence of OA is 5.343 per 100 thousand individuals and its incidence is 430.4 per 100 thousand individuals in the Middle East and North African countries [4]. In a study conducted in Europe, it was emphasized that OA-related drug use was 46.9% and that patients with pain were disconnected from work life, which brought a burden to the healthcare system [5]. In a study conducted on very elderly individuals in the Ural-Altay region, it was stated that the prevalence was 27.7 and 13.8% of the patients were using medication [6]. As society ages and quality life expectancy increases, OA has begun to impose a threat on life quality of patients causing social and psychological disturbances [1]. In a study conducted in 2014, it was stated that OA is expected to increase from 26.6 to 29.5% in individuals over the age of 45 in 2032 [7]. Knee involvement in OA is seen in 85% of patients. The European OA report published in 2022 emphasizes that the direct cost of the disease reached 7.2 billion Euros and the indirect cost reached 4.4 billion Euros [8]. Intensive studies are being carried out to prevent pain, movement limitation and deformity in OA patients with multifactorial and multiple genetic backgrounds by early recognition and to monitor efficiency of treatments [9]. One of these studies is the European Union Cooperation in Science and Technology (COST) Action CA21110 entitled as "Building an open European Network on OsteoArthritis research" that aims to coordinate interdisciplinary research and technological developments to improve life quality of OA patients by creating a common structure in which patients, physicians and researchers from academia and industry can participate [10].

In OA, in addition to articular joint cartilage, synovium, subchondral bone, joint capsule and almost all periarticular ligaments are also affected [11]. Since articular

Table 1 Common characteristics of OA

• A cyclical disease with multiple phenotypes, multiple etiologies, heterogeneous and low inflammatory
• Advanced age, female gender, black race, heredity, increased body mass index, trauma, inactivity [12]
• Morbidity and mortality [13]
• Meniscus tear, intra-articular or periarticular fracture and ligament injury or joint dislocation in secondary OA [16]
• The symptoms are cyclically progressing pain, swelling, limitation of movement and deformity [17]
• Decrease in the thickness of the articular cartilage, deterioration of its molecular structure and change in its microarchitecture: change in chondroitin-keratan sulfate ratios: chondro-senescence
• Degeneration (catabolic degradation) through the cycle of fibrillation, destruction and regeneration (anabolic construction) in articular joint cartilage
• Thinning and sclerosis on the articular surface
• Subchondral cyst formation and painful osteophytes with inflammation in the subchondral bone
• Thickening of the joint capsule, synovitis and decrease in joint fluid: changes in metabolites
• Meniscus degeneration and displacement from the joint space
• Skeletal muscle atrophy: sarcopenia
• Wearing of tendons and ligaments: deterioration of balance due to laxity and increased risk of falling

joint cartilage lacks blood vessels, it carries out its metabolic activity through the joint fluid synthesized by the synovium. For this reason, metabolites of inflammation, degeneration and regeneration cycles of the articular joint cartilage are expected to concentrate in the synovial joint fluid [12]. It is therefore important to evaluate the synovial joint fluid with omic technology. Inflammatory diseases affecting articular joint cartilage and the destruction and production of molecules caused by environmental factors should mostly concentrate in the synovial joint fluid (Fig. 1) [12]. The International OA Research Association (www.oarsi.org) stated that the disease progresses with extracellular matrix degradation and cell stress after injury in the joints, inflammation pathways are activated and metabolic deterioration in the joint causes anatomical and physiological changes [13, 14]. It was suggested in 1975 that OA begins with inflammation (Fig. 1) [15]. Therefore, inflammatory markers are expected to concentrate in synovial joint fluid during the cyclic processes of OA [18]. Although there are ELISA and RT-qPCR devices and products that study the degradation and production metabolites of articular cartilage from blood, these are currently used only for research purposes [19].

2 Background Information: Sports-Related Early Osteoarthritis

Anterior cruciate ligament (ACL) rupture predominantly affects young and active patients, with a majority under 30 years old. Anterior cruciate ligament ruptures are more common in females than males and the incidence is 0.81 per 1.000 people. Various factors including anatomical, hormonal and biomechanical factors related to ACL ruptures were evaluated. The fact that about 50% of individuals with ACL rupture develop OA within 5–15 years is alarming as the incidence of ACL ruptures in younger generations is increasing. Meniscus tears also have a bimodal distribution. Acute tears occur in young and active patients, while chronic degenerative tears affect older individuals, with a higher incidence in males. The overall incidence of meniscus tears is 0.7 per 1.000 patient-years. After arthroscopic partial meniscectomy, 38% of patients with medial meniscectomy and 24% of patients with lateral meniscectomy experience radiographic OA changes within 4.5 years. Furthermore, 71% of patients develop radiographic OA within 21 years following total meniscectomy. Shoulder instability primarily affects young and active patients, with males being more prone to it. The incidence varies with 0.08 per 1.000 patient-years in the general population and a higher rate of 1.7 per 1.000 patient per year in active military recruits. Twenty-six percent of patients who require surgery for shoulder instability exhibit radiographic signs of OA. Patellar instability is also common in young and active patients, which is often linked to anatomical predisposition. Patellar instability is also prevalent in females. The incidence ranges from 0.058 per 1.000 patient-years in the general population to 0.29 per 1.000 patient-years in the 10–17 age group. Fifty-two percent of patients who experience a single acute joint

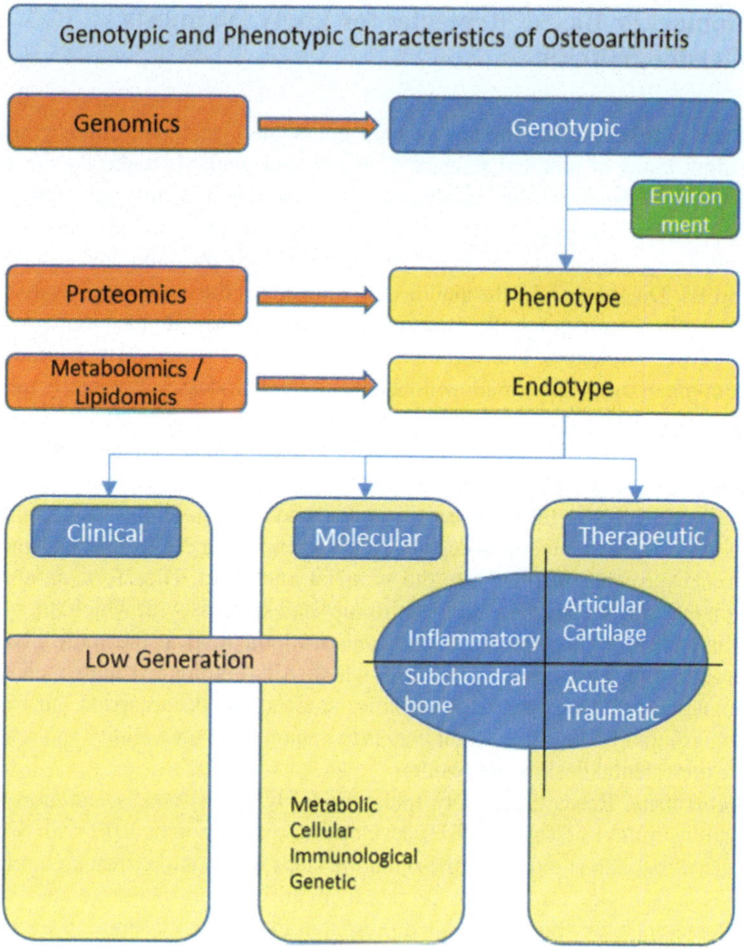

Fig. 1 Intersection regions of omics analyzes with genotypic and phenotypic features of OA

dislocation suffer from long-term movement limitations that may also cause early OA. Athletes are particularly susceptible to ankle instability, with an incidence of 0.1 episode per 1.000 patient-days. Thirteen percent of patients with chronic ankle instability develop radiographic signs of OA during a 20-year follow-up [20]. Sport-related early OA therefore relates to a sudden decrease in quality of life in a younger population, which also limits treatment options.

3 Biomarker-Based Strategies for Early Diagnosis of Osteoarthritis

Diagnosis of diseases, monitoring of their course and treatment efficacy has been the leading focus of medical science. Conventional methods make the early diagnosis of the disease and the monitoring of its progression difficult. Advances in laboratory technologies and chromatographic methods provided the opportunity to examine clinical pictures at the molecular level; offered time, labor and cost effective alternatives. Diagnosis of arthropathies, monitoring of the course and evaluation of treatment efficacy are generally made with the patient through questionnaires based on qualitative parameters. The development of an analytical method with which quantitative comparisons can be made will benefit clinical processes, especially treatment planning. Although synovial joint fluid is a lubricant for the joint area, it is a transfer medium for the cartilage extracellular matrix and molecules released during the vascular circulation. Structures such as the synovium, meniscus, ligament, and joint capsule express protein through secretion into the synovial joint fluid. Anabolites and catabolites formed as a result of regeneration/degeneration events for bone and cartilage homeostasis also diffuse into the synovial joint fluid. Therefore, an abnormal situation that can be experienced in the tissues and systems with which the synovial joint fluid interacts is reflected in the synovial joint fluid at the molecular level. The examination of synovial joint fluid, which is the biological matrix containing the biomarkers that are the focus of omics studies, has been carried out by many working groups. What is important here is to demonstrate the validity and reliability of the experimental design and results.

Osteoarthritis Research Society International (OARSI) and Foundation for the National Institutes of Health (FNIH) developed a consortium in 2012 with two main goals. The first is to standardize the scientific steps of biomarker research related to the development of treatments for OA and to identify factors in age, gender, physical activity and episodic changes of the disease that may affect baseline measurements [21, 22]. The second goal was to identify highly validated biomarkers that are predictive of clinical outcomes of OA [23]. In line with the decisions of the consortium, a series of studies evaluating the role of specific biomarkers in the pathogenesis of OA and the metabolic impact of clinical treatments have been concluded [23–25]. Drawing on the work of FNIH and OARSI, the biomarker validation process requires the precise identification of biomarkers based on sensitivity, specificity, false positive probability, and false negative probability. This issue was also brought to the agenda at the World OA Congress held between 17 and March 20, 2023, and it was decided to continue the work on this issue meticulously [26].

Early diagnosis of OA may lead to preventive measures of the disease and/or condition. The ultimate aim is to modify the natural course of the condition and slow down its development. Planning the treatment with the most appropriate approach and monitoring its success is equally important (Fig. 2) [18]. Current OA diagnostics heavily relies on patient history, physical examination and radiology [21]. These methodologies have limitations as the progression of the disease and/or the success of

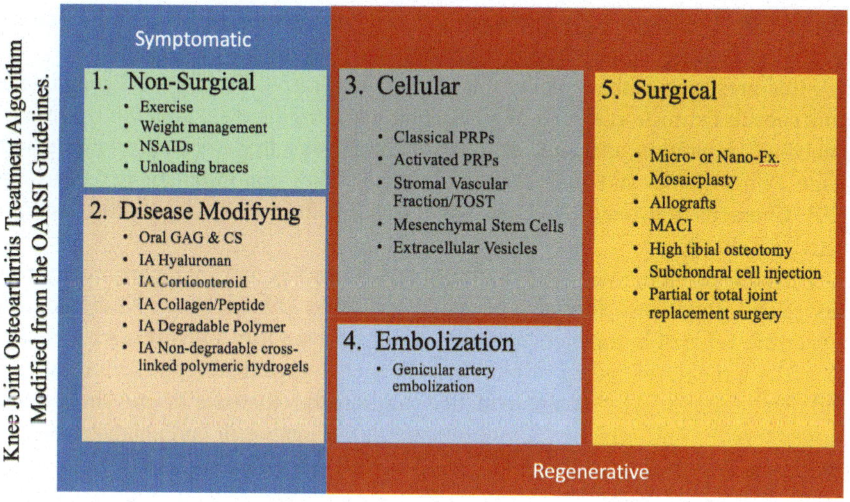

Fig. 2 Knee OA treatment algorithm

the treatment cannot be adequately monitored with such methods [27]. Radiological findings and symptoms during physical examination may not always overlap with each other. These examination methods are also limited in the early diagnosis of OA. With magnetic resonance, articular cartilage can be imaged and mapped in certain centers with specific softwares [28]. This approach, which requires trained medical staff and a developed infrastructure, is also expensive and may vary from assessment to assessment. The fact that OA does not cause homogeneous destruction on the joint surface additionally limits magnetic resonance methods.

There is a need to develop a reliable and economical biomarker-based diagnostic kit that can provide rapid results at the bedside of the patient in order to diagnose OA as early as possible. Monitoring the success of treatments also seems to be essential (Fig. 2). This issue was also addressed within the scope of the European Union's OActive project (https://www.oactive.eu/) in 2018 and was supported with a budget of nearly five million Euros. Biomarkers have some advantages. These can be stated as the early diagnosis of OA, distinguishing between subgroups and phenotypes of the disease and determining the severity of OA [27, 28].

3.1 Synovial Joint Fluid: A Biomarkers Bank

Synovial joint fluid is a biological structure that is a physiological lubricant for regular function. It is also a transfer medium for the joint cartilage molecules. Synovial joint fluid is produced by the synovium as an ultrafiltrate of blood plasma. It mainly consists of lubricin, hyaluronan, collagenases, proteinase and prostaglandins. Synovial joint fluid production is carried out by fibroblast-like type B synovial cells,

Physiological changes in the volume and content of synovial joint fluid occur in response to trauma, inflammation, and bacterial, fungal, or viral penetration.

Structures such as the joint cartilage, synovium, meniscus, labrum, ligament and joint capsule express proteins through secretion into the synovial joint fluid. Synovial joint fluid aspiration and analysis are very important in patients presenting with acute painful joints and suspected inflammatory and non-inflammatory effusions [29]. Characteristic features of the synovial joint fluid in different situations are listed in Table 2.

The role of biomarkers in OA diagnosis, prognosis and treatment monitoring has gained importance in recent years. While many studies have investigated the diagnostic value of biomarkers to determine disease severity, other researchers have focused on the development of new OA treatments targeting biomarker production and regulation [30, 42]. Recent evidence suggests the existence of presymptomatic biomarkers for OA, which is asymptomatic in the long term and characterized by progression of tissue destruction [13, 32, 33]. Early identification of OA biomarkers could allow faster treatment than currently used tools and prevent or delay cartilage and bone damage resulting in clinical disability if accurately located [31]. Biomarkers also provide the opportunity to monitor the response of therapies such as corticosteroids, hyaluronic acid and cellular therapies such as platelet-rich plasma (PRP), which are now commonly used in the treatment of OA [13]. One of the examples that can be given systematically in this field is ORBIT (Orthobiologics Initiative) Consensus established by ESSKA (European Society for Sports Traumatology, Knee Surgery and Arthroscopy), which focuses on two fundamental issues: cell-based therapy and non-transfusional hemo-components. Biomarker-focused research could be used to monitor the success of new treatment modalities. In terms of its relationship to osteoarthritis, biomarkers may be related to various tissue or physiological processes. It is important to focus on cartilage-derived biomarkers for osteoarthritis, which is mainly characterized by a degenerative outcome in cartilage tissue. These biomarkers may be degenerative, regenerative or inflammatory in origin. Cartilage-derived biomarkers are listed in Table 3.

Table 2 Synovial joint fluid features [34]

Specification	Normal	Non-inflammatory	Inflammatory	Septic
Color	Clear	Yellow	Yellow, opalescent	Yellow—Green
Clarity	Transparent	Transparent	Translucent-opaque	Opaque
Viscosity	High	High	Low	Variable
WBC/mm3	< 200	200–2000	2000–10.000	> 50.000
Polymorphonuclear leukocytes (PMNs)	< 25%	~ 25%	> 50%	> 75%
Culture	Negative	Negative	Negative	Often positive
Glucose (mg/dL)	Nearly = to plasma	Nearly = to plasma	~ 25, lower than plasma	< 25, even lower than plasma

Table 3 Joint cartilage biomarkers classified according to their molecular properties [35]

Formation Biomarkers: PIICP, PIIANP, PIIBNP, CPII, IGF-1, FGF, **HA, Coll 2**
Degradation Biomarkers: CTX-II, C2C, CIIM, HELIX-II, Coll 2–1 NO2, **COMP**, CS846, Aggrecan Epitopes (ARGS, TEGE, FFGV), Fibulin-3 Epitopes (Fib3- 1, Fib3-2, Fib3-3), TIINE, Gly-Gal-Pyd, MMP2, **MMP3**, MMP9, MMP13, TIMP-2, TIMP-3, TIMP-4, ADAMTS-4, ADAMTS-5
Inflammatory Biomarkers: IL-1 β, IL-1Ra, TNF, TNF-Rs, IL-2, IL-4, **IL-6, IL-8, IL-10,** IL-15, IL-17, IL-18, CXCL-10, PGE2, IL-8/CXCL-8, GROa/CXCL-1, MCP-1/CCL-2, RANTES/CCL-5, MIP-1α/CCL-3, MIP-1β/CCL-4, CCL2, CCL3, CCL, CCL5, CCL7, CCL21, VEGF, 15-HETE, CRP, CRPM, Resistin, Leptin, Adioponektin, ApoA1, MPO

PIICP: Procollagen II C-Terminal ProPeptide, PIIANP: Type IIA Collagen N-Propeptide, PIIBNP: The type II collagen N-propeptide, CPII: Type II collagen propeptide, IGF-1: Insulin-like growth factor-1, FGF: Fibroblast growth factors, HA: Hyaluronic acid, Coll 2: Collagen type II, CTX-II: Crosslinked C-telopeptides of type II collagen, C2C: Neoepitope of type II collagen, CIIM: Matrix metalloproteinase-derived fragment of type II collagen, HELIX-II: Helical peptide of type II collagen, Coll 2–1 NO2: Nitrated form of triple helical region of type II collagen, COMP: Cartilage oligomeric matrix protein, CS846: Chondroitin sulfate 846 epitope, TIINE: Type II collagen neoepitope, Gly-Gal-Pyd: Glucosyl-galactosyl-pyridinoline, MMP: Matrix metalloproteinase, TIMP: Tissue inhibitor of metalloprotease, ADAMTS: A disintegrin and metalloproteinase with thrombospondin motifs, IL: Interleukin, CXCL-10: CXC motif chemokine ligand, PGE2: Prostaglandin E2, GROa: Growth-regulated oncogene-alpha, MCP-1: Monocyte Chemoattractant Protein-1, CCL: The chemokine (C–C motif) ligand, RANTES: Raised on activation normal T-cell derived and secreted, MIP: Macrophage inhibitory protein, VEGF: Vascular endothelial growth factor, 15-HETE: 15-Hydroxy-eicosatetraenoic acid, CRP: C-reactive protein, CRPM: C-reactive protein metabolite, ApoA1: Apolipoprotein, MPO: myeloperoxidase

3.1.1 Advantages and Disadvantages of Current Synovial Joint Fluid Biomarkers

An ideal biomarker should meet the requirements of accessibility, repeatability, reproducibility, specificity and clinical relevance. Biomarkers studied have been valued at the level of which and to what extent they meet these requirements. Biomarkers, which predominantly focus on human-derived protein groups may cause high costs due to their validation studies. Biomarkers associated with OA are also at the intersection of other inflammatory diseases. Biomarkers such as COMP, CTX-II, C2C are therefore more meaningful in this respect. When studies on biomarkers are examined, it is seen that most of the studies (Supporting Information—Table S4) are produced by global mapping or untargeted mass spectrometry methods. Conducting targeted mass spectrometry studies with the data obtained from these studies is important for future studies. Another limitation of the studies is the evaluation made from a one-way perspective. Creating an analyte extract with affinity chromatography in order to focus on a specific target group and being able to approach many physiological processes holistically with the multi omic approach are the unique aspects of such studies.

3.2 Affinity Chromatography Supported Strategies

Affinity chromatography allows selective separation of the target molecule from a complex matrix. While different targets are focused on in many biological matrices, phosphopeptide studies have not been found especially in synovial joint fluid. We focus on the intergroup phosphopeptide profiles of the synovial joint fluid that can be used in OA diagnosis and treatment planning. Phosphorylation regulates proliferation, signal transduction and apoptosis at the cellular level. Phosphorylation is an event known to occur in one-third of the human proteome. Phosphoproteins can also be found at different sites in the human proteome, but in lesser amounts compared to other proteins. Only a few percent of all proteins are phosphated. Phosphorylation is involved in most signal transduction mechanisms and cellular activities. Therefore, studies in the field of phosphoproteomics, in which phosphoproteins are examined, are very important as a branch of proteomics. Since important diseases such as cancer and Alzheimer's are related to the mechanisms in which proteins such as kinases and phosphatases are involved, treatment methods are tried to be developed for diseases arising from the deterioration of such mechanisms, by considering them as target molecules. To date, some inhibitor compounds and antibodies related to these target molecules have started to take place in clinical use as a result of the progress of the last ten years. The combination of very low and very high amounts of proteins in complex biological samples such as the synovial joint fluid in mass spectroscopy analysis creates a significant problem in establishing the protein profile accurately. The phosphoproteomics study workflow is schematized in Fig. 3. Enrichment step, in which very low amounts of phosphoproteins are generally enriched at the peptide level, is therefore of critical importance. Various enrichment strategies developed based on the chemical properties of the phosphate group are grouped as immobilized metal affinity chromatography (IMAC), metal oxide affinity chromatography (MOAC), immune affinity and chemical derivatization [36–42]. In the IMAC method, which is one of the oldest enrichment strategies, Fe^{+3}, Al^{+3} and Ga^{+3} metal ions are generally chelated in a matrix environment [49]. This method is based on the principle that these positively charged metal ions capture negatively charged phosphate groups. Similar principles are used in the MOAC method where strong acidic buffers are used but metal oxides such as TiO_2, ZrO_2, Fe_3O_4 are covalently bonded to phosphopeptides. Many modified forms of the IMAC and MOAC methods have been developed using different acid types and buffers at different pH [43–47]. Enrichment with TiO_2 is widely used in global phosphoproteomic studies. A commercial sorbent TiO was previously used for the synovium in a study [48]. When the sorbents synthesized for phosphopeptide enrichment were examined (Supporting Information—Table S1), it was seen that these studies were mainly used for optimization in terms of sensitivity, reusability and selectivity. Studies used in plasma, saliva, cell lysate, cell lines and standard phosphoproteins were listed. Phosphopeptide enrichment studies for the synovial joint fluid were less-evaluated. We therefore aimed to evaluate the information obtained from the isolated biomolecules for the diagnosis of OA. The phosphorylated macromolecules (phosphoproteins) and their digested forms (phosphopeptides) were evaluated by metal oxide affinity chromatography from the synovial joint fluids of volunteers/patients for this purpose. The chemical structure of all the isolated molecules were identified using the MALDI-TOF and/or the LC–MS/MS techniques. The usability of the information generated based on the identified chemical structures isolated by their molecules were evaluated.

Phosphoproteins can be isolated from complex matrix environments such as the synovial joint fluid by affinity chromatography. It is predicted that the MOAC-IMAC affinity sorbent we synthesized was adsorbed by targets for its synergistic effect and its branched structure. Treating the matrix in this way before omics studies is a target-oriented approach that eliminates randomness. Comparison of phosphoprotein/phosphopeptide profiles between groups formed part of the multi-omic approach. Lipidomic, proteomic and metabolomic analysis can be performed and targeted validation studies can also be carried out.

3.3 *Omics-Based Strategies*

"Omics" technologies were developed to analyze DNA, RNA, proteins and amino acids, lipids and similar metabolites on a large scale. These technologies were used to study the pathogenesis of OA, monitor the natural course of the disease and/or decide for the accuracy of treatment modalities during the last decade [47]. Metabolomic studies in OA were summarized by a research group in 2018 [50]. Articles published in the last five years on synovial joint fluid were examined and summarized (Supporting Information—Table S5). Briefly, a total of 38 publications were examined and it was noteworthy that the studies generally included a singular approach and contained a small number of samples, where the experimental and control groups could not be well designed due to limiting reasons. It was observed that the validity of the method was not examined in most of the studies (Supporting Information—Table S2).

3.3.1 Metabolomics

Metabolomic studies reveal changes at the metabolome level, allowing the identification of metabolic pathways affected by the disease state [51, 52]. These studies provide instant information about the metabolome level, allowing the phenotype to be examined at the metabolome level in addition to the genotype elucidated at the genome level. Today, research at the metabolome level is commonly carried out with liquid chromatography—mass spectrometry (LC–MS) based devices [53]. Untargeted metabolomics allows us to screen a wide range of metabolites in the sample and detect changes at the metabolome level semi-quantitatively between groups. Targeted metabolomics, on the other hand, are studies in which predetermined metabolites or pathways are targeted and these metabolites are detected quantitatively [51, 54]. A study has been published that examined the metabolomic fingerprint in joint fluid and

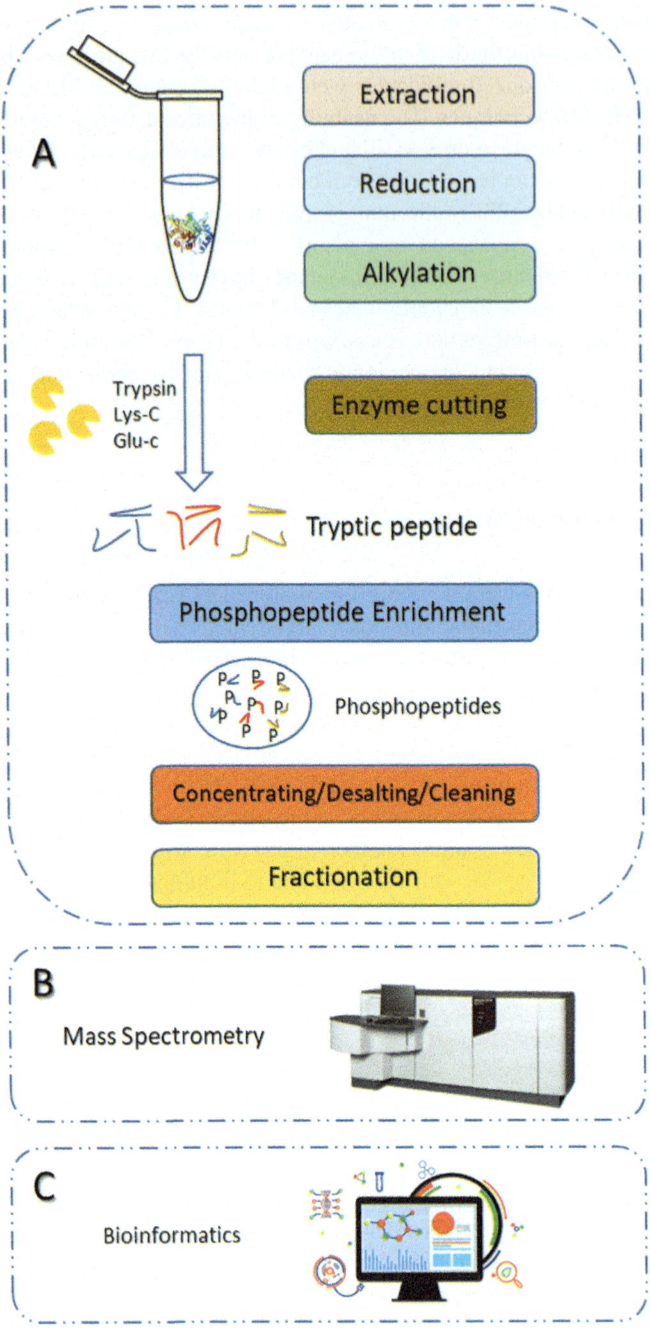

Fig. 3 Schematic representation of phosphoproteomic analysis workflow

subpatellar fat tissue using an untargeted method [55]. A study in 2021 performed metabolomic analysis in the synovial joint fluid of 30 rheumatoid arthritis patients and revealed disorders in pyrimidine and purine metabolisms, fatty acid biosynthesis and unsaturated fatty acid synthesis [56].

3.3.2 Lipidomics

Lipids of the synovial joint fluid have not been widely studied yet [57]. The presence of the subpatellar fat tissue in the structure of the knee joint raises the possibility that lipid molecules may also be present in the joint fluid [58–60]. Lipids are classified depending on head group and type of linkage between aliphatic chains and head group [61]. The MAPS consortium maintains a database that includes a list of lipid classes and subclasses and abbreviations [62]. There are studies in the literature showing that there are changes in phospholipid amounts with aging. Phospholipids have been added to the literature as potential biomarkers in many diseases. Changes in phosphoprotein amounts were evaluated as inflammation markers, cardiovascular diseases, cancer and other diseases [63–67]. Lipidomics, an approach that emerged in 2023 to study the metabolism of the cellular lipidome is the study of lipids using the principles and techniques of analytical chemistry [65, 68, 69]. New developments in the field of analytical chemistry, improvements in mass spectrometry (MS) and the incredible development in information technology have made this emerging discipline rapidly applicable [70]. Today, lipidomics is the most common method to determine lipid biomarkers in the study of diseases [71]. With lipidomic analysis, cellular metabolism can be examined by measuring changes in individual lipid classes, subclasses and molecular species that reflect metabolic differences [72]. Since lipid metabolism is well known today, changes in lipid ratios can simultaneously reveal activity at several enzymatic levels or variations in gene expression levels [73]. Good processing and interpretation of lipidomic data can yield results almost equivalent to enzymatic assays, which can identify molecular mechanisms responsible for disease pathogenesis [74, 75]. In a recent study, extracellular exosomes obtained from joint fluid of patients with OA were examined and it was stated that lipidomic profiles may differ depending on the isolation method [76].

3.3.3 Proteomics

Proteomics-based studies focus on the identification of proteins and peptides. The composition of extracellular matrices (ECMs) from various normal and pathological tissues has been characterized using a combination of ECM-enrichment methods and mass spectrometry. As anticipated, these research findings have demonstrated the existence of ECM proteins that are widely distributed throughout various tissues, while also identifying proteins that are exclusive to certain tissues [77]. For instance, the Önnerfjord's group conducted a study that demonstrated variations in ECM compositions among different cartilages [78]. In addition, they demonstrated that the composition of articular cartilage exhibits spatial variation using a combination of targeted mass spectrometry and quantitative analysis. They found that proteins were distributed in three layers of articular cartilage. While asporin, tenascin-C, thrombospondin-4 and perlecan were in the superficial layer of articular cartilage, mimecan and thrombospondin-1 were found in the intermediate layer. The most enriched proteins in the deeper layers of articular cartilage were aggrecan, osteomodulin and chondroadherin [78]. The LC–MS/MS technique has been employed to observe the distinct proteomes of various joint tissues, cells, as well as bodily fluids in both healthy individuals and those with OA [79]. After more than a decade of proteomic studies in OA [80], these approaches have contributed to a better understanding of the pathogenesis of the disease and the identification of new protein markers. The most recent descriptive studies in this field have focused on elucidating the molecular composition of cartilage and the disease-related changes that occur in this tissue [81]. A recent study conducted in 2022 examined the proteomic analysis of exosomes isolated from synovial joint fluid of patients with rheumatoid arthritis, spondyloarthritis, gout and OA by using LC–MS [81]. It was emphasized that the proteomic analysis of exosomes isolated from joint fluid showed differences between diseases and that they could be used in the development of biomarkers in future. In a proteomic study conducted in patients with anterior cruciate ligament rupture, it was shown that the levels of glutathione peroxidase 1 and plasticin 3 were increased in patients with isolated anterior cruciate ligament rupture compared to patients with anterior cruciate ligament rupture with meniscus tear. It has been stated that leptin and glucose 6 phosphate dehydrogenase increase in patients with accompanying bone marrow edema [82]. It was found that glutathione peroxidase 1 level decreased in patients with bone marrow edema in the tibial plateau. It was observed that matrix metalloproteinase 9 and G6PD values were higher in male patients than in females. A study evaluating the proteomic study and ELISA results together has not been conducted. In another recent study published in 2023, the joint fluid of patients with OA was compared with the joint fluid of normal individuals and it was revealed that the proteomic profile was different in women [83]. Proteomics research in osteoarthritis (OA) involves the comprehensive analysis of numerous proteins, enabling the identification of functional elements within chondrocytes and cartilage tissue. This approach also allows for the detection of dynamic alterations in the protein composition resulting from interactions between proteins and post-translational modifications (PTMs) [84]. As can be seen from current publications, the use of omics-based technologies in the investigation of OA and similar musculoskeletal system injuries or diseases is increasing. When comparing the proteomic approach to metabolomics and genomics, it becomes evident that the former possesses some advantages. Notably, the proteomic technique is capable of providing a snapshot of the patient's status at a certain moment in time [79]. Furthermore, it exhibits greater stability in comparison to metabolites. The addition of metabolomic and lipidomic studies to proteomic and phosphoproteomic studies and the realization of targeted proteomic studies will provide an important basis for future research.

3.3.4 Limitations of Omics

In general, omic studies provide a broad spectrum of important biological information at different biomolecule levels. However, there may also be important limiting situations. First of all, no analytical method can analyze the entire metabolome or proteome in its entirety. In these cases, either additional purification in the sample preparation steps or modifications to the analytical method may be required [85]. In addition to analytical measurements, situations such as sample size, natural errors due to sample collection and storage, and the involvement of the same biomolecule in more than one pathway should be carefully monitored in clinical studies. The phenotype may differ depending on the person's lifestyle and daily habits as well as the disease. When evaluating experimental designs and the results obtained, it should be taken into account whether the difference arises from the factor in question in the experiment or from the difference in phenotype between individuals [86]. Metabolomics and lipidomics databases are expanding and our knowledge about biological pathways and their functions is increasing day by day, but it should not be forgotten that there are still many points waiting to be elucidated at the molecular level.

4 Our Study Design and Progress: IMAC-Assisted Multi-omics Approach

4.1 Synovial Joint Fluid Aspiration

Arthrocentesis, the procedure for collecting synovial joint fluid through joint space aspiration, necessitates meticulous attention to sterile conditions and should be performed by a physician well-versed in the relevant anatomy. In this process, sterility is not only vital to prevent infection transmission but also crucial for accurate synovial joint fluid analysis.

Initiating the procedure involves prepping and draping the designated site using antiseptics such as alcohol, betadine and/or chlorhexidine in the standard sterile fashion. A local anesthetic should be applied to relieve cutaneous and subcutaneous localized pain. At least 10 to 15 min should be allowed between sterilization and local anesthetic applications. In this study, we used both cutaneous and subcutaneous anesthetics for the comfort of our patients. Following this, a large gauge needle, typically 18 gauge, connected to a 50 cc syringe, was carefully inserted into the knee joint through the antero-lateral portal. Anatomical landmarks were difficult to find in obese and highly degenerated joints and marking these landmarks before sterilization was considered as a safer approach in such patients. Negative pressure in the syringe was avoided and samples that contained blood macroscopically were not used for evaluations. At least 10 ml of synovial joint fluid was collected from each patient. The collected synovial joint fluid was then transferred to 2 ml sterile specimen collection tubes, sealed and transported to −20 °C for freezing. Post-procedure, the antiseptic was cleaned off the skin, the site was dried and pressure with a soft dressing was applied for local hemostasis. Additional compression and cold application was not utilized in any of our patients.

Synovial joint fluid obtained from 12 patients with joint swelling after an acute knee injury (Control Group) and OA (Experimental Group) over 65 years of age will be evaluated. The ion suppression effect is commonly observed in LC/MS studies as a natural consequence of injections without enrichment or pre-separation. Phosphopeptide enrichment methods eliminate this phenomenon and allow better results to be observed in this peptide group.

4.2 Synthesis of IMAC Sorbent

Importance of advanced technologies in the diagnosis and monitoring treatment outcomes of chronic diseases is progressing. In osteoarthritis (OA), proteins such as kinases and phosphatases are involved. These are targeted molecules for diagnosis and treatment. Irregularities in the mechanisms related to protein phosphorylation in OA trigger signaling pathways, causing cartilage destruction and inflammation. In the first phase of the proposed interdisciplinary and translational project, we aimed to synthesize and characterize a new IMAC sorbent to be used in the phosphoproteomic studies of the synovial joint fluid. This new sorbent should be reusable and present a higher adsorption capacity compared to commercial ones. For the purpose of affinity sorbent synthesis, poly(MAA-co-EDMA)-based microspheres were used as mold material. Monodisperse, porous and high surface area SiO_2 microspheres were obtained. PEI (polyethyleneimine) was used as the interface layer to immobilize Ti(IV) ions. Thanks to the branched chain structure of PEI, it is aimed to immobilize more Ti(IV) ions. As a final step for the synthesis of SiO_2@PEI@Ti(IV), Ti(IV) ion is attached to interact with phosphopeptides (Fig. 4). Various enrichment strategies developed based on the chemical properties of the phosphate group; immobilized metal affinity chromatography (IMAC), metal oxide affinity chromatography (MOAC), immune affinity and chemical derivatization. In the IMAC method, which is one of the oldest enrichment strategies, generally Fe(III), Al(III) and Ga(III) metal ions are chelated in a matrix environment and it is based on the fact that these positively charged metal ions capture negatively charged phosphate groups. Phosphoproteins can be attached to the surface by using the specific interaction between the phosphate groups in the phosphoproteins and the immobilized Ti(IV) cations. Proteins are polypeptide molecules. Therefore, proteins must be digested into peptide fractions for targeted detection and quantification studies at the peptide level. For this purpose, trypsin enzyme, which is also found in the digestive system and has the ability to digest proteins, was used. To our knowledge, affinity sorbent was synthesized for the first time in synovial joint fluid for phosphopeptidomic analysis.

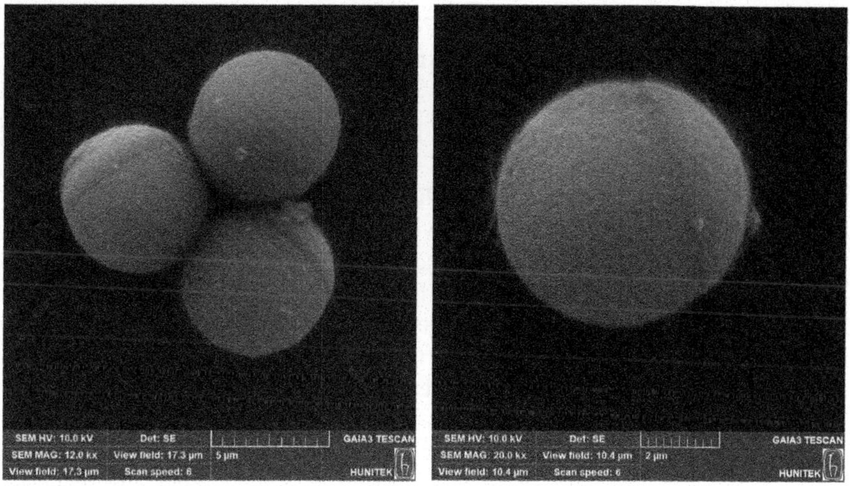

Fig. 4 Scanning electron microscope images of SiO_2@PEI@Ti(IV) sorbents for phosphopeptide enrichment

4.3 Phosphopeptide Enrichment Process

The phosphoprotein enrichment method with Ti(IV)-IMAC generally consists of three steps. In the first step, the sample is loaded. For this, phosphoprotein solution is prepared and the prepared solution is interacted with Ti(IV)-IMAC material until the adsorption equilibrium is reached. In the second step, non-specific peptides and proteins are removed. With the help of a centrifuge, the material is separated from the solution and the aqueous phase containing the unadsorbed proteins is separated from the Ti(IV)–IMAC material. In the last step, the elution of the captured phosphoproteins is provided. Protein-loaded Ti(IV)–IMAC material is taken into the elution medium and the proteins are separated from the surface. Phosphopeptide enrichment was carried out by interacting the peptide solution obtained after tryptic digestion with affinity sorbents. In order to measure the phosphopeptide isolation performance of the synthesized affinity sorbent, known phosphoproteins α-casein and β-casein were used. Tryptic digestion and phosphopeptide enrichment processes have been validated in this step. Standard phosphoproteins and eluents obtained by tryptic digestion and phosphopeptide enrichment from synovial joint fluid were used for mass spectrometry analysis.

4.4 Investigation of Phosphopeptide Enrichment and Detection Capability Using LC–MS/MS

In order to confirm the phosphopeptide isolation performance of the synthesized particle, the particle whose performance was previously evaluated was selected as the control group. a, b and c refer to eluants obtained from SiO_2@PEI@Ti(IV), SiO_2@PDA@Ti(IV) and digested synovial joint fluid eluant, respectively. When Fig. 5c is examined, it is seen that there are many non-specific peaks in the sample obtained by digesting the synovial joint fluid. Figure 5b shows the peaks detected after phosphopeptide enrichment with the particle considered as the control group. Figure 5c shows the phosphopeptide enrichment and isolation performance of the new particle we synthesized. As can be seen, preliminary studies are promising about the new particle.

Figure 6 shows the number of phosphopeptide sequences detected from the eluants that we named a, b and c. It was observed that 28, 26 and 13 phosphopeptide sequences were detected, respectively. Detailed tables are included in the supporting information (Table S1, S2, S3). As a result of the studies performed with LC–MS/MS, the phosphopeptide isolation performance of the affinity sorbent was demonstrated chromatographically.

4.5 Expectations from the Future

The new sorbent will facilitate the analysis of synovial joint fluid in proteomic, metabolomic and lipidomic studies. Within the scope of the project, besides the phosphoproteomic study, metabolomic and lipidomic analysis will be performed in the synovial joint fluid. Metabolites of physiological and pathological production and destruction of articular cartilage are concentrated in the synovial joint fluid. Phosphorylation regulates proliferation, signal transduction and apoptosis at the cellular level. With this new technology the phenotypic, clinical, molecular and therapeutic endotypes of OA will be evaluated in future studies. Omics specific to inflammation, metabolic, cellular and immunological phases of OA will be mapped. As a result, the phosphopeptide enrichment performance of the new sorbent will be demonstrated by mass spectrometry. Within the scope of omic studies, proteomic, phosphoproteomic, metabolomic and lipidomic investigations of synovial joint fluid will be performed by mass spectrometry. It is aimed to pave the way for the determination of new diagnosis and treatment molecules by evaluating the data with bioinformatics methods to be obtained from omic data and ELISA studies together.

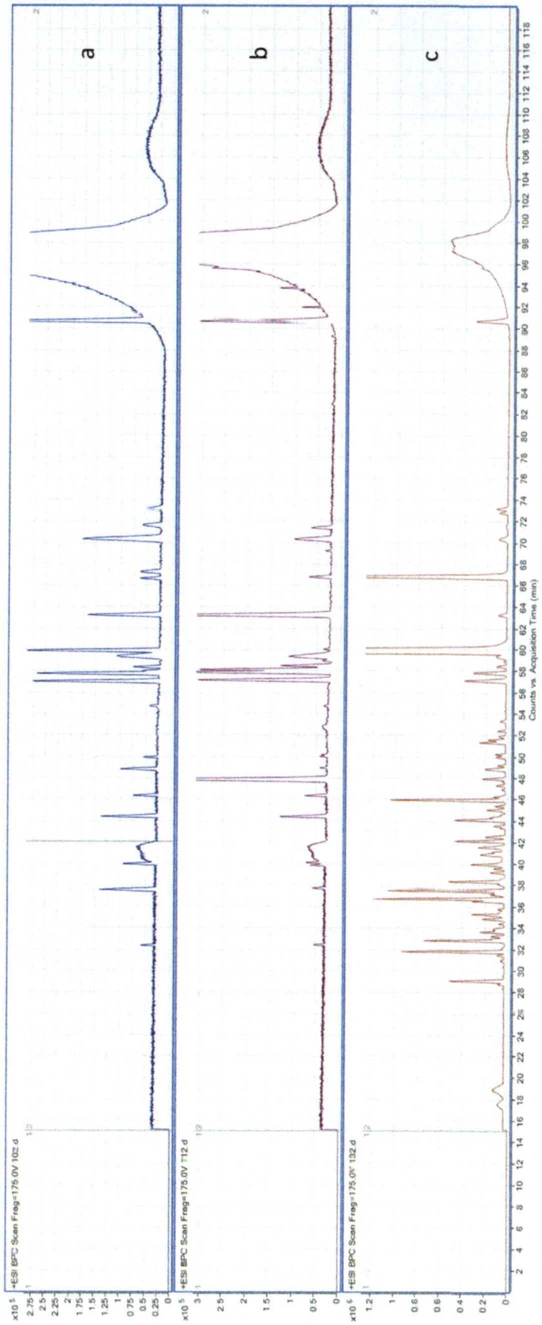

Fig. 5 Demonstration of the phosphopeptide determination performance of synovial fluid before and after phosphopeptide enrichment with IMAC sorbents by mass spectrometry

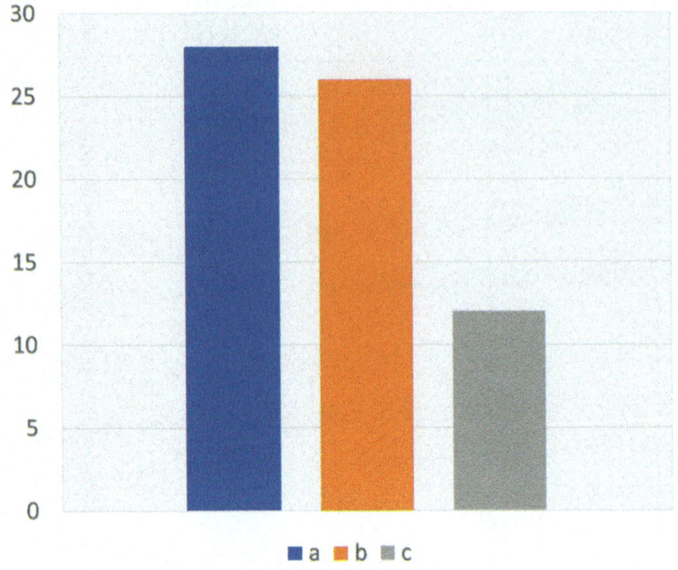

Fig. 6 Number of detected phosphopeptide sequence

5 Conclusion and Clinical Implication

Diagnosis, treatment and monitoring of the course of OA depends on several factors including genetics, epigenetics and patients pheno- and endo-types. Diagnosis is currently based on patient history, physical examination and medical imaging. Research on biomarkers is promising and genomics, proteomics and metabolomics are trending areas. Controlling variables and validation studies however may gain importance in ensuring the validity and reliability of the results. Clinical, molecular and therapeutic endotypes may also define the stage of the condition. Inflammation to some extent determines degenerative and regenerative conditions of the articular joint cartilage, subchondral bone and all other joint tissues. The synovium itself and the synovial joint fluid are important tissues and components that deserve attention. Disease modifying biomedical and medical treatments mostly focus on suppressing the inflammation and preventing tissue degeneration. Cellular approaches may stimulate regeneration through their paracrine effects. Monitoring the disease and treatment modalities using omic technologies is therefore becoming more important. Optimizing the processes of collecting synovial joint fluid samples and preparing them for analysis may prevent differences between studies. Including correlation results during the execution of omics studies can prevent false positive outcomes. In our study, we evaluated publications on metabolomic, lipidomic and proteomic data of the synovial joint fluid while examining phosphopeptides. Early findings were displayed and the methodology of sample obtainment was verified. Mapping metabolites in accordance with inflammation, degeneration and regeneration tools were optimized using ELISA. The creation of point-of-care diagnosis and monitoring modalities will be developed once the relationship of the detected biomarkers in OA will be demonstrated. Various disease modifying technologies including biomaterials and cells are frequently used in the clinic currently. Clinical applications are far ahead of basic science knowledge. Information on clinical outcomes is mostly based on personal experience and low level of evidence studies. Limited number of cases with short follow-up may need molecular and metabolomic explanations. Quantitatively evaluating the synovial joint fluid may give further information on whether we should continue our treatment and/or change it with the other modalities especially in young OA patients.

Acknowledgements The Scientific and Technological Research Council of Türkiye (TÜBİTAK) granted this study in relevance to the EU COST CA21110—"Building an open European Network on OsteoArthritis research (NetwOArk)" Action. Part of this study was presented at the Cells & Extracellular Templates (CET)—2nd International Conference in Rome between June 8 and 9, 2023, and 3rd Regenerative Medicine Orthopaedic Summit (RMOS) in İstanbul between November 30 and December 2, 2023.

Author Contributions Serhat Aladağ, Feza Korkusuz, Petek Korkusuz, Mustafa Çelebier, Ömür Çelikbıçak and Ali Tuncel created the idea. All authors defined the concept of the study. İdil Yet, Serhat Aladağ, Feza Korkusuz, Petek Korkusuz, Mustafa Çelebier, Ömür Çelikbıçak and Ali Tuncel designed the study. Serhat Aladağ, İlayda Demirdiş, Burcu Gökçal Kapucu and Ali Tuncel worked for the development of the IMAC sorbents section. Batuhan Erhan Aktaş and Feza Korkusuz prepared the synovial joint fluid acquisition and storage sections. İlayda Demirdiş, Bilge Başak Fidan, Ozan Kaplan, Melis Şardan Ekiz, Ömür Çelikbıçak and Mustafa Çelebier prepared the LC–MS/MS analysis section. Merve Gizer and Petek Korkusuz wrote the ELISA experiment sections. Serhat Aladağ and Feza Korkusuz prepared the original draft. All authors equally contributed to the final writing and reviewing of the manuscript. All authors contributed to the final version of the tables, supplementary materials, figures and editing of the manuscript. All authors approved the submitted version.

Funding This study was supported by Scientific and Technological Research Council of Türkiye (TUBITAK) under the Grant Number 223S5094. The authors thank to TUBITAK for their supports. This work also was supported by Research Fund of the Hacettepe University under the Project Number: 21455.

Declarations

Conflict of Interest This study is based on the PhD Thesis of Serhat Aladağ entitled "*Development of Monodisperse-Mesoporous SiO_2 Microspheres as a New Metal Oxide Affinity Chromatography (MOAC) Sorbent for Phospholipid Enrichment in Synovial Joint Fluid*". Feza Korkusuz MD and Ali Tuncel are active members of the Turkish Academy of Sciences (TÜBA). The authors declare that they have no known competing financial interests or personal relationships that could have appeared to influence the work reported in this paper.

Ethics Committee Permission and Approval Ethics committee approval was received for this study from the Hacettepe University Health Sciences Research Ethics Committee (Approval no and date:2023/09-44-19.12.2023, Research no: SBA 23/262).

Abbreviations

ACL	Anterior cruciate ligament
ECMs	Extracellular matrices
ELISA	Enzyme-Linked Immunosorbent Assay
FNIH	Foundation for the National Institutes of Health
IMAC	Immobilized metal affinity chromatography
MOAC	Metal oxide affinity chromatography
MALDI-TOF	Matrix-Assisted Laser Desorption/Ionization time-of-flight
LC–MS/MS	Liquid chromatography–mass spectrometry
MAPS	Metabolites and Pathways Strategy
OA	Osteoarthritis
OARSI	Osteoarthritis Research Society International
PDA	Polydopamine
PEI	Polyethyleneimine
PTMs	Post-translational modifications
TÜBA	Turkish Academy of Sciences
TÜBİTAK	The Scientific and Technological Research Council of Türkiye

References

1. Quicke JG, Conaghan PG, Corp N, Peat G (2022) Osteoarthritis year in review 2021: epidemiology & therapy. Osteoarthritis Cartilage 30(2):196–206. https://doi.org/10.1016/j.joca.2021.10.003
2. Cui A, Li H, Wang D, Zhong J, Chen Y, Lu H (2020) Global, regional prevalence, incidence and risk factors of knee osteoarthritis in population-based studies. EClinicalMedicine 29–30:100587. https://doi.org/10.1016/j.eclinm.2020.100587
3. Yahaya I, Wright T, Babatunde OO, Corp N, Helliwell T, Dikomitis L, Mallen CD (2021) Prevalence of osteoarthritis in lower middle- and low-income countries: a systematic review and meta-analysis. Rheumatol Int 41(7):1221–1231. https://doi.org/10.1007/s00296-021-04838-y
4. Shamekh A, Alizadeh M, Nejadghaderi SA, Sullman MJM, Kaufman JS, Collins GS, Kolahi AA, Safiri S (2022) The burden of osteoarthritis in the Middle East and North Africa Region from 1990 to 2019. Front Med (Lausanne). 9:881391. https://doi.org/10.3389/fmed.2022.881391
5. Kingsbury SR, Gross HJ, Isherwood G, Conaghan PG (2014) Osteoarthritis in Europe: impact on health status, work productivity and use of pharmacotherapies in five European countries. Rheumatology (Oxford) 53(5):937–947. https://doi.org/10.1093/rheumatology/ket463
6. Bikbov MM, Kazakbaeva GM, Gilmanshin TR, Zainullin RM, Rakhimova EM, Fakhretdinova AA, Tuliakova AM, Rusakova IA, Panda-Jonas S, Nuriev IF, Zaynetdinov AF, Zinnatullin AA, Arslangareeva II, Gizzatov AV, Bolshakova NI, Safiullina KR, Jonas JB (2022) Prevalence and associated factors of osteoarthritis in the Ural eye and medical study and the Ural very old study. Sci Rep 12(1):12607. https://doi.org/10.1038/s41598-022-16925-6
7. Turkiewicz A, Petersson IF, Björk J, Hawker G, Dahlberg LE, Lohmander LS, Englund M (2014) Current and future impact of osteoarthritis on health care: a population-based study with

projections to year 2032. Osteoarthritis Cartil 22(11):1826–1832. https://doi.org/10.1016/j.joca.2014.07.015
8. Ortho Consulting Group (2023) Research Report: Osteoarthritis in the EU. https://orthocg.com/wp-content/uploads/2022/06/Osteoarthritis-in-the-EU-Research-Report-June-2022.pdf
9. Ratneswaran A, Kapoor M (2021) Osteoarthritis year in review: genetics, genomics, epigenetics. Osteoarthritis Cartil 29(2):151–160. https://doi.org/10.1016/j.joca.2020.11.003
10. European Cooperation in Science and Technology (COST). https://www.cost.eu/about/about-cost/
11. O'Neill TW, Felson DT (2018) Mechanisms of Osteoarthritis (OA) Pain. Curr Osteoporos Rep 16(5):611–616. https://doi.org/10.1007/s11914-018-0477-1
12. Boffa A, Merli G, Andriolo L, Lattermann C, Salzmann GM, Filardo G (2021) Synovial fluid biomarkers in knee osteoarthritis: a systematic review and quantitative evaluation using BIPEDs criteria. Cartilage 13(1_suppl):82S-103S. https://doi.org/10.1177/1947603520942941
13. Katz JN, Arant KR, Loeser RF (2021) Diagnosis and treatment of hip and knee osteoarthritis: a review. JAMA 325(6):568–578. https://doi.org/10.1001/jama.2020.22171
14. Ehrlich GE (1975) Inflammation as an important feature of osteoarthritis. J Am Med Assoc 232:157–159
15. Wang LJ, Zeng N, Yan ZP, Li JT, Ni GX (2020) Post-traumatic osteoarthritis following ACL injury. Arthritis Res Ther 22(1):57. https://doi.org/10.1186/s13075-020-02156-5
16. Klement MR, Sharkey PF (2019) The significance of osteoarthritis-associated bone marrow lesions in the knee. J Am Acad Orthop Surg 27(20):752–759. https://doi.org/10.5435/JAAOS-D-18-00267
17. Singh A (2021) Biosensor for electrochemical detection of biomarkers of osteoarthritis. Osteoarthritis Cartil 29:S165–S166. https://doi.org/10.1016/j.joca.2021.02.230
18. Hu SL, Chang AC, Huang CC, Tsai CH, Lin CC, Tang CH (2017) Myostatin promotes interleukin-1β expression in rheumatoid arthritis synovial fibroblasts through inhibition of miR-21-5p. Front Immunol 8:1747. https://doi.org/10.3389/fimmu.2017.01747
19. Grässel S, Muschter D (2020) Recent advances in the treatment of osteoarthritis. F1000Res 9:F1000 Faculty Rev-325. https://doi.org/10.12688/f1000research.22115.1
20. Carbone A, Rodeo S (2017) Review of current understanding of post-traumatic osteoarthritis resulting from sports injuries. J Orthop Res 35(3):397–405. https://doi.org/10.1002/jor.23341
21. Dinçel YM (2018) Value of biomarkers in osteoarthritis. In: Osteoarthritis biomarkers and treatments; *IntechOpen*: London, UK, vol 5
22. Kraus VB, Burnett B, Coindreau J, Cottrell S, Eyre D, Gendreau M, Gardiner J, Garnero P, Hardin J, Henrotin Y, Heinegård D, Ko A, Lohmander LS, Matthews G, Menetski J, Moskowitz R, Persiani S, Poole AR, Rousseau JC, Todman M (2011) Application of biomarkers in the development of drugs intended for the treatment of osteoarthritis. Osteoarthritis Cartil 19(5):515–542. https://doi.org/10.1016/j.joca.2010.08.019
23. Kraus VB, Collins JE, Hargrove D, Losina E, Nevitt M, Katz JN, Wang SX, Sandell LJ, Hoffmann SC, Hunter DJ (2017) Predictive validity of biochemical biomarkers in knee osteoarthritis: data from the FNIH OA Biomarkers Consortium. Ann Rheum Dis 76(1):186–195. https://doi.org/10.1136/annrheumdis-2016-209252
24. Deveza LA, Kraus VB, Collins JE, Guermazi A, Roemer FW, Bowes M, Nevitt MC, Ladel C, Hunter DJ (2017) Association between biochemical markers of bone turnover and bone changes on imaging: data from the osteoarthritis initiative. Arthritis Care Res (Hoboken). 69(8):1179–1191. https://doi.org/10.1002/acr.23121
25. Ruiz-Romero C, Rego-Perez I, Blanco FJ (2018) What did we learn from 'omics' studies in osteoarthritis. Curr Opin Rheumatol 30(1):114–120. https://doi.org/10.1097/BOR.0000000000000460
26. Pre-Congress Workshop 3: The STEpUP OA Consortium—Large Scale Agnostic Molecular Analysis of Synovial Fluid to Search for Endotypes in OA. https://2023.oarsi.org/program/sessions/pre-congress-workshop-3-stepup-oa-consortium-%E2%80%93-large-scale-agnostic-molecular Accessed Date 15 July 2023

27. Karanfil Y, Babayeva N, Dönmez G, Diren HB, Eryılmaz M, Doral MN, Korkusuz F (2019) Thirty minutes of running exercise decreases T2 signal intensity but not thickness of the knee joint cartilage: a 3.0-t magnetic resonance imaging study. Cartilage 10(4):444–450. https://doi.org/10.1177/1947603518770246
28. Lana JF, Rodrigues B (2019) Osteoarthritis as a chronic inflammatory disease: a review of the inflammatory markers. IntechOpen. https://doi.org/10.5772/intechopen.82565
29. Seidman AJ, Limaiem F (2023) Synovial fluid analysis. [Updated 2023 May 1]. In: StatPearls [Internet]. Treasure Island (FL): StatPearls Publishing. Available from: https://www.ncbi.nlm.nih.gov/books/NBK537114/
30. Watt FE (2018) Osteoarthritis biomarkers: year in review. Osteoarthritis Cartil 26(3):312–318. https://doi.org/10.1016/j.joca.2017.10.016
31. Kapoor M, Mahomed N (2015) Osteoarthritis: pathogenesis, diagnosis, available treatments, drug safety, regenerative and precision medicine. Adis. London, UK, pp 1–278
32. Toumi H, Marija M (2019) Osteoarthritis biomarkers and treatments. IntechOpen. https://doi.org/10.5772/intechopen.76957
33. Thudium CS, Löfvall H, Karsdal MA, Bay-Jensen AC, Bihlet AR (2019) Protein biomarkers associated with pain mechanisms in osteoarthritis. J Proteomics 190:55–66. https://doi.org/10.1016/j.jprot.2018.04.030
34. García-De La Torre I (2006) Advances in the management of septic arthritis. Infect Dis Clin North Am 20(4):773–788. https://doi.org/10.1016/j.idc.2006.09.005
35. Nguyen LT, Sharma AR, Chakraborty C, Saibaba B, Ahn ME, Lee SS (2017) Review of prospects of biological fluid biomarkers in osteoarthritis. Int J Mol Sci 18(3):601. https://doi.org/10.3390/ijms18030601
36. Batalha IL, Lowe CR, Roque AC (2012) Platforms for enrichment of phosphorylated proteins and peptides in proteomics. Trends Biotechnol 30(2):100–110. https://doi.org/10.1016/j.tibtech.2011.07.004
37. Boersema PJ, Foong LY, Ding VM, Lemeer S, van Breukelen B, Philp R, Boekhorst J, Snel B, den Hertog J, Choo AB, Heck AJ (2010) In-depth qualitative and quantitative profiling of tyrosine phosphorylation using a combination of phosphopeptide immunoaffinity purification and stable isotope dimethyl labeling. Mol Cell Proteomics 9(1):84–99. https://doi.org/10.1074/mcp.M900291-MCP200
38. Thaler F, Valsasina B, Baldi R, Xie J, Stewart A, Isacchi A, Kalisz HM, Rusconi L (2003) A new approach to phosphoserine and phosphothreonine analysis in peptides and proteins: chemical modification, enrichment via solid-phase reversible binding, and analysis by mass. Anal Bioanal Chem 376(3):366–373. https://doi.org/10.1007/s00216-003-1919-9
39. Tsumoto H, Ra M, Samejima K, Taguchi R, Kohda K (2008) Chemical derivatization of peptides containing phosphorylated serine/threonine for efficient ionization and quantification in matrix-assisted laser desorption/ionization time-of-flight mass spectrometry. Rapid Commun Mass Spectrom 22(7):965–972. https://doi.org/10.1002/rcm.3451
40. Negroni L, Claverol S, Rosenbaum J, Chevet E, Bonneu M, Schmitter JM (2012) Comparison of IMAC and MOAC for phosphopeptide enrichment by column chromatography. J Chromatogr B Analyt Technol Biomed Life Sci 891–892:109–112. https://doi.org/10.1016/j.jchromb.2012.02.028
41. Aryal UK, Ross AR (2010) Enrichment and analysis of phosphopeptides under different experimental conditions using titanium dioxide affinity chromatography and mass spectrometry. Rapid Commun Mass Spectrom 24(2):219–231. https://doi.org/10.1002/rcm.4377
42. Ruprecht B, Koch H, Medard G, Mundt M, Kuster B, Lemeer S (2015) Comprehensive and reproducible phosphopeptide enrichment using iron immobilized metal ion affinity chromatography (Fe-IMAC) columns. Mol Cell Proteomics 14(1):205–215. https://doi.org/10.1074/mcp.M114.043109
43. Tsai CF, Hsu CC, Hung JN, Wang YT, Choong WK, Zeng MY, Lin PY, Hong RW, Sung TY, Chen YJ (2014) Sequential phosphoproteomic enrichment through complementary metal-directed immobilized metal ion affinity chromatography. Anal Chem 86(1):685–693. https://doi.org/10.1021/ac4031175

44. Larsen MR, Thingholm TE, Jensen ON, Roepstorff P, Jørgensen TJ (2005) Highly selective enrichment of phosphorylated peptides from peptide mixtures using titanium dioxide microcolumns. Mol Cell Proteomics 4(7):873–886. https://doi.org/10.1074/mcp.T500007-MCP200
45. Wan H, Yan J, Yu L, Zhang X, Xue X, Li X, Liang X (2010) Zirconia layer coated mesoporous silica microspheres used for highly specific phosphopeptide enrichment. Talanta 82(5):1701–1707. https://doi.org/10.1016/j.talanta.2010.07.050
46. Blacken GR, Volný M, Diener M, Jackson KE, Ranjitkar P, Maly DJ, Tureček F (2009) Reactive landing of gas-phase ions as a tool for the fabrication of metal oxide surfaces for in situ phosphopeptide enrichment. J Am Soc Mass Spectrom 20:915–926. https://doi.org/10.1016/j.jasms.2009.01.006
47. Chen SY, Juang YM, Chien MW, Li KI, Yu CS, Lai CC (2011) Magnetic iron oxide nanoparticle enrichment of phosphopeptides on a radiate microstructure MALDI chip. Analyst 136:4454–4459. https://doi.org/10.1039/C1AN15334J
48. Tang S, Deng S, Guo J, Chen X, Zhang W, Cui Y, Luo Y, Yan Z, He QY, Shen S, Wang T (2018) Deep coverage tissue and cellular proteomics revealed IL-1β can independently induce the secretion of TNF-associated proteins from human synoviocytes. J Immunol 200(2):821–833. https://doi.org/10.4049/jimmunol.1700480
49. Salimi K, Usta DD, Çelikbıçak Ö, Pınar A, Salih B, Tuncel A (2017) Highly selective enrichment of phosphopeptides by titanium (IV) attached monodisperse-porous poly(vinylphosphonic acid-co-ethylene dimethacrylate) microspheres. J Chromatogr A 1496:9–19. https://doi.org/10.1016/j.chroma.2017.03.052
50. Szponder T, Latalski M, Danielewicz A, Krać K, Kozera A, Drzewiecka B, Nguyen Ngoc D, Dobko D, Wessely-Szponder J (2022) Osteoarthritis: pathogenesis, animal models, and new regenerative therapies. J Clin Med 12(1):5. https://doi.org/10.3390/jcm12010005
51. Bujak R, Struck-Lewicka W, Markuszewski MJ, Kaliszan R (2015) Metabolomics for laboratory diagnostics. J Pharm Biomed Anal 113:108–120. https://doi.org/10.1016/j.jpba.2014.12.017
52. Sari ZB (2020) Metabolomik Çalışmaların Dünü, Bugünü ve Yarını. Türkiye Klinikleri Tip Bilimleri Dergisi. 40(4):480. https://doi.org/10.5336/medsci.2020-79692
53. Zhang XW, Li QH, Xu ZD, Dou JJ (2020) Mass spectrometry-based metabolomics in health and medical science: a systematic review. RSC Adv 10(6):3092–3104. https://doi.org/10.1039/c9ra08985c
54. Zhang A, Sun H, Yan G, Wang P, Wang X (2015) Metabolomics for biomarker discovery: moving to the clinic. Biomed Res Int 2015:354671. https://doi.org/10.1155/2015/354671
55. Kishimoto K, Urade R, Ogawa T, Moriyama T (2001) Nondestructive quantification of neutral lipids by thin-layer chromatography and laser-fluorescent scanning: suitable methods for "lipidome" analysis. Biochem Biophys Res Commun 281(3):657–662. https://doi.org/10.1006/bbrc.2001.4404
56. Wang H, Fang K, Wang J, Chang X (2021) Metabolomic analysis of synovial fluids from rheumatoid arthritis patients using quasi-targeted liquid chromatography-mass spectrometry/mass spectrometry. Clin Exp Rheumatol 39(6):1307–1315. https://doi.org/10.55563/clinexprheumatol/s5jzzf
57. Nieminen P, Hämäläinen W, Savinainen J, Lehtonen M, Lehtiniemi S, Rinta-Paavola J, Lehenkari P, Kääriäinen T, Joukainen A, Kröger H, Paakkonen T, Mustonen AM (2022) Metabolomics of synovial fluid and infrapatellar fat pad in patients with osteoarthritis or rheumatoid arthritis. Inflammation 45(3):1101–1117. https://doi.org/10.1007/s10753-021-01604-x
58. Yetukuri L, Katajamaa M, Medina-Gomez G, Seppänen-Laakso T, Vidal-Puig A, Oresic M (2007) Bioinformatics strategies for lipidomics analysis: characterization of obesity related hepatic steatosis. BMC Syst Biol 1:12. https://doi.org/10.1186/1752-0509-1-12
59. Han X, Jiang X (2009) A review of lipidomic technologies applicable to sphingolipidomics and their relevant applications. Eur J Lipid Sci Technol 111(1):39–52. https://doi.org/10.1002/ejlt.200800117

60. Yang K, Cheng H, Gross RW, Han X (2009) Automated lipid identification and quantification by multidimensional mass spectrometry-based shotgun lipidomics. Anal Chem 81(11):4356–4368. https://doi.org/10.1021/ac900241u
61. Han X (2016) Lipidomics for studying metabolism. Nat Rev Endocrinol 12(11):668–679. https://doi.org/10.1038/nrendo.2016.98
62. Fahy E, Subramaniam S, Brown HA, Glass CK, Merrill AH, Murphy RC, Raetz CR, Russell DW, Seyama Y, Shaw W, Shimizu T, Spener F, van Meer G, VanNieuwenhze MS, White SH, Witztum JL, Dennis EA (2005) A comprehensive classification system for lipids. J Lipid Res 46(5):839–861. https://doi.org/10.1194/jlr.E400004-JLR200
63. Nitschke P, Lodge S, Kimhofer T, Masuda R, Bong SH, Hall D, Schäfer H, Spraul M, Pompe N, Diercks T, Bernardo-Seisdedos G, Mato JM, Millet O, Susic D, Henry A, El-Omar EM, Holmes E, Lindon JC, Nicholson JK, Wist J (2022) J-Edited diffusional proton nuclear magnetic resonance spectroscopic measurement of glycoprotein and supramolecular phospholipid biomarkers of inflammation in human serum. Anal Chem 94(2):1333–1341. https://doi.org/10.1021/acs.analchem.1c04576
64. Fraley AE, Schwartz GG, Olsson AG, Kinlay S, Szarek M, Rifai N, Libby P, Ganz P, Witztum JL, Tsimikas S (2009) Relationship of oxidized phospholipids and biomarkers of oxidized low-density lipoprotein with cardiovascular risk factors, inflammatory biomarkers, and effect of statin therapy in patients with acute coronary syndromes: results from the MIRACL (Myocardial Ischemia Reduction With Aggressive Cholesterol Lowering) trial. J Am Coll Cardiol 53(23):2186–2196. https://doi.org/10.1016/j.jacc.2009.02.041
65. Ashraf MZ, Kar NS, Podrez EA (2009) Oxidized phospholipids: biomarker for cardiovascular diseases. Int J Biochem Cell Biol 41(6):1241–1244. https://doi.org/10.1016/j.biocel.2008.11.002
66. Glunde K, Serkova NJ (2006) Therapeutic targets and biomarkers identified in cancer choline phospholipid metabolism. Pharmacogenomics 7(7):1109–1123. https://doi.org/10.2217/14622416.7.7.1109
67. Philippova M, Resink T, Erne P, Bochkov V (2014) Oxidised phospholipids as biomarkers in human disease. Swiss Med Wkly 144:w14037. https://doi.org/10.4414/smw.2014.14037
68. Han X, Gross RW (2003) J Lipid Res 44(6):1071–1079. https://doi.org/10.1194/jlr.R300004-JLR200
69. Lee SH, Williams MV, DuBois RN, Blair IA (2003) Targeted lipidomics using electron capture atmospheric pressure chemical ionization mass spectrometry. Rapid Commun Mass Spectrom 17(19):2168–2176. https://doi.org/10.1002/rcm.1170
70. Hu T, Zhang JL (2018) Mass-spectrometry-based lipidomics. J Sep Sci 41(1):351–372. https://doi.org/10.1002/jssc.201700709
71. Lv J, Zhang L, Yan F, Wang X (2018) Clinical lipidomics: a new way to diagnose human diseases. Clin Transl Med 7(1):12. https://doi.org/10.1186/s40169-018-0190-9
72. Zhao YY, Cheng XL, Lin RC (2014) Lipidomics applications for discovering biomarkers of diseases in clinical chemistry. Int Rev Cell Mol Biol 313:1–26. https://doi.org/10.1016/B978-0-12-800177-6.00001-3
73. Brown AJ, Coates HW, Sharpe LJ (2021) Chapter 10—cholesterol synthesis. In: Ridgway ND, McLeod RS (eds) Biochemistry of lipids, lipoproteins and membranes. 7. Amsterdam, Elsevier, pp 317–55
74. Kisluk J, Ciborowski M, Niemira M, Kretowski A, Niklinski J (2014) Proteomics biomarkers for non-small cell lung cancer. J Pharm Biomed Anal 101:40–49. https://doi.org/10.1016/j.jpba.2014.07.038
75. Kvasnička A, Najdekr L, Dobešová D, Piskláková B, Ivanovová E, Friedecký D (2023) Clinical lipidomics in the era of the big data. Clin Chem Lab Med 61(4):587–598. https://doi.org/10.1515/cclm-2022-1105
76. Chen D, Yu H, Zhang Y, Huang Y (2022) Metabolomic analysis of extracellular vesicles from human synovial fluids. Microchem J 177:107257, ISSN 0026–265X
77. Naba A, Clauser KR, Ding H, Whittaker CA, Carr SA, Hynes RO (2016) The extracellular matrix: tools and insights for the "omics" era. Matrix Biol. https://doi.org/10.1016/j.matbio.2015.06.003

78. Önnerfjord P, Khabut A, Reinholt FP, Svensson O, Heinegård D (2012) Quantitative proteomic analysis of eight cartilaginous tissues reveals characteristic differences as well as similarities between subgroups. J Biol Chem. https://doi.org/10.1074/jbc.M111.298968
79. Tardif G, Paré F, Gotti C, Roux-Dalvai F, Droit A, Zhai G, Sun G, Fahmi H, Pelletier J-P, Martel-Pelletier J (2022) Mass spectrometry-based proteomics identify novel serum osteoarthritis biomarkers. Arthritis Res Ther. https://doi.org/10.1186/s13075-022-02801-1
80. Ali N, Turkiewicz A, Hughes V, Folkesson E, Tjörnstand J, Neuman P, Önnerfjord P, Englund M (2022) Proteomics profiling of human synovial fluid suggests increased protein interplay in early-osteoarthritis (OA) that is lost in late-stage OA. Mol Cell Proteomics 21(3):100200. https://doi.org/10.1016/j.mcpro.2022.100200
81. Huang Y, Liu Y, Huang Q, Sun S, Ji Z, Huang L, Li Z, Huang X, Deng W, Li T (2022) TMT-based quantitative proteomics analysis of synovial fluid-derived exosomes in inflammatory arthritis. Front Immunol 13:800902. https://doi.org/10.3389/fimmu.2022.800902
82. Brophy RH, Cai L, Zhang Q, Townsend RR, Rai MF (2022) Proteomic profile analysis of synovial fluid in patients with anterior cruciate ligament tears. Am J Sports Med 50(11):2935–2943. https://doi.org/10.1177/03635465221112652
83. Muller PR, Lee TJ, Zhi W, Kumar S, Vyavahare S, Sharma A, Kumar V, Isales CM, Hunter M, Fulzele S (2023) Proteomic analysis of female synovial fluid to identify novel biomarkers for osteoarthritis. Life (Basel). 13(3):605. https://doi.org/10.3390/life13030605
84. Trachana V, Mourmoura E, Papathanasiou I, Tsezou A (2019) Understanding the role of chondrocytes in osteoarthritis: utilizing proteomics. Expert Rev Proteomics. https://doi.org/10.1080/14789450.2019.1571918
85. Kohler I, Verhoeven A, Derks RJ, Giera M (2016) Analytical pitfalls and challenges in clinical metabolomics. Bioanalysis 8(14):1509–1532. https://doi.org/10.4155/bio-2016-0090
86. Cui L, Lu H, Lee YH (2018) Challenges and emergent solutions for LC-MS/MS based untargeted metabolomics in diseases. Mass Spectrom Rev 37(6):772–792. https://doi.org/10.1002/mas.21562
87. Salimi K, Usta DD, Çelikbıçak Ö, Pinar A, Salih B, Tuncel A (2017) Ti(IV) carrying polydopamine-coated, monodisperse-porous SiO_2 microspheres with stable magnetic properties for highly selective enrichment of phosphopeptides. Colloids Surf B Biointerfaces 153:280–290. https://doi.org/10.1016/j.colsurfb.2017.02.028
88. Zhao L, Qin H, Hu Z, Zhang Y, Wu R, Zou H (2012) A poly(ethylene glycol)-brush decorated magnetic polymer for highly specific enrichment of phosphopeptides. Chem Sci 3:2828–2838. https://doi.org/10.1039/C2SC20363D
89. Yan Y, Zheng Z, Deng C, Li Y, Zhang X, Yang P (2013) Hydrophilic polydopamine-coated graphene for metal ion immobilization as a novel immobilized metal ion affinity chromatography platform for phosphoproteome analysis. Anal Chem 85(18):8483–8487. https://doi.org/10.1021/ac401668e
90. Zhu X, Gu J, Yang J, Wang Z, Li Y, Zhao L, Zhao W, Shi J (2015) Zr-based metal–organic frameworks for specific and size-selective enrichment of phosphopeptides with simultaneous exclusion of proteins. J Mater Chem B 3(20):4242–4248. https://doi.org/10.1039/c5tb00113g
91. Xie Y, Deng C (2016) Highly efficient enrichment of phosphopeptides by a magnetic lanthanide metal-organic framework. Talanta 159:1–6. https://doi.org/10.1016/j.talanta.2016.05.075
92. Piovesana S, Capriotti AL, Cavaliere C, Ferraris F, Samperi R, Ventura S, Laganà A (2016) Phosphopeptide enrichment: Development of magnetic solid phase extraction method based on polydopamine coating and Ti(4+)-IMAC. Anal Chim Acta 909:67–74. https://doi.org/10.1016/j.aca.2016.01.008
93. Salimi K, Usta DD, Çelikbıçak Ö, Pinar A, Salih B, Tuncel A (2017) Ti(IV) carrying polydopamine-coated, monodisperse-porous SiO2 microspheres with stable magnetic properties for highly selective enrichment of phosphopeptides. Colloids Surf B Biointerf 2017153:280–290. https://doi.org/10.1016/j.colsurfb.2017.02.028
94. Liu Q, Sun N, Gao M, Deng, C (2018) Magnetic binary metalorganic framework as a novel affinity probe for highly selective capture of endogenous phosphopeptides. ACS Publications. J Contribut. https://doi.org/10.1021/acssuschemeng.8b00023.s001

95. Arribas Diez I, Govender I, Naicker P, Stoychev S, Jordaan J, Jensen ON (2021) Zirconium(IV)-IMAC revisited: improved performance and phosphoproteome coverage by magnetic microparticles for phosphopeptide affinity enrichment. J Proteome Res 20(1):453–462. https://doi.org/10.1021/acs.jproteome.0c00508
96. Xu Z, Wu Y, Wu H, Sun N, Deng C (2021) Hydrophilic polydopamine-derived mesoporous channels for loading Ti(IV) ions for salivary phosphoproteome research. Anal Chim Acta 1146:53–60. https://doi.org/10.1016/j.aca.2020.12.038
97. Gao L, Tao J, Qi L, Jiang X, Shi H, Liu Y, Di B, Wang Y, Yan F (2022) Synthesis of a metal oxide affinity chromatography magnetic mesoporous nanomaterial and development of a one-step selective phosphopeptide enrichment strategy for analysis of phosphorylated proteins. Anal Chim Acta 1195:339430. https://doi.org/10.1016/j.aca.2022.339430
98. Zhai G, Yang L, Luo Q, Wu K, Zhao Y, Wang F (2022) Serum phosphopeptide profiling for colorectal cancer diagnosis using liquid chromatography-mass spectrometry. Rapid Commun Mass Spectrom 36(15):e9316. https://doi.org/10.1002/rcm.9316
99. Mickiewicz B, Kelly JJ, Ludwig TE, Weljie AM, Wiley JP, Schmidt TA, Vogel HJ (2015) Metabolic analysis of knee synovial fluid as a potential diagnostic approach for osteoarthritis. J Orthop Res 33(11):1631–1638. https://doi.org/10.1002/jor.22949
100. Jin W, Woo DC, Jahng GH (2016) In vivo H1 MR spectroscopy using 3 Tesla to investigate the metabolic profiles of joint fluids in different types of knee diseases. J Appl Clin Med Phys 17(2):561–572. https://doi.org/10.1120/jacmp.v17i2.6144
101. Kim S, Hwang J, Kim J, Ahn JK, Cha HS, Kim KH (2017) Metabolite profiles of synovial fluid change with the radiographic severity of knee osteoarthritis. Joint Bone Spine 84(5):605–610. https://doi.org/10.1016/j.jbspin.2016.05.018
102. Zheng K, Shen N, Chen H, Ni S, Zhang T, Hu M, Wang J, Sun L, Yang X (2017) Global and targeted metabolomics of synovial fluid discovers special osteoarthritis metabolites. J Orthop Res 35(9):1973–1981. https://doi.org/10.1002/jor.23482
103. Anderson JR, Chokesuwattanaskul S, Phelan MM, Welting TJM, Lian LY, Peffers MJ, Wright HL (2018) 1H NMR metabolomics identifies underlying inflammatory pathology in osteoarthritis and rheumatoid arthritis synovial joints. J Proteome Res 17(11):3780–3790. https://doi.org/10.1021/acs.jproteome.8b00455
104. Carlson AK, Rawle RA, Adams E, Greenwood MC, Bothner B, June RK (2018) Application of global metabolomic profiling of synovial fluid for osteoarthritis biomarkers. Biochem Biophys Res Commun 499(2):182–188. https://doi.org/10.1016/j.bbrc.2018.03.117
105. Carlson AK, Rawle RA, Wallace CW, Brooks EG, Adams E, Greenwood MC, Olmer M, Lotz MK, Bothner B, June RK (2019) Characterization of synovial fluid metabolomic phenotypes of cartilage morphological changes associated with osteoarthritis. Osteoarthritis Cartilage 27(8):1174–1184. https://doi.org/10.1016/j.joca.2019.04.007
106. Kim S, Hwang J, Kim J, Lee SH, Cheong YE, Lee S, Kim KH, Cha HS (2022) Metabolic discrimination of synovial fluid between rheumatoid arthritis and osteoarthritis using gas chromatography/time-of-flight mass spectrometry. Metabolomics 18(7):48. https://doi.org/10.1007/s11306-022-01893-9
107. Jaggard MKJ, Boulangé CL, Graça G, Akhbari P, Vaghela U, Bhattacharya R, Williams HRT, Lindon JC, Gupte CM (2023) The effect of liquid-liquid extraction on metabolite detection and analysis using NMR spectroscopy in human synovial fluid. J Pharm Biomed Anal 226:115254. https://doi.org/10.1016/j.jpba.2023.115254
108. Kosinska MK, Liebisch G, Lochnit G, Wilhelm J, Klein H, Kaesser U, Lasczkowski G, Rickert M, Schmitz G, Steinmeyer J (2013) A lipidomic study of phospholipid classes and species in human synovial fluid. Arthritis Rheum 65(9):2323–2333. https://doi.org/10.1002/art.38053
109. Kosinska MK, Liebisch G, Lochnit G, Wilhelm J, Klein H, Kaesser U, Lasczkowski G, Rickert M, Schmitz G, Steinmeyer J (2014) Sphingolipids in human synovial fluid–a lipidomic study. PLoS ONE 9(3):e91769. https://doi.org/10.1371/journal.pone.0091769
110. Kosinska MK, Ludwig TE, Liebisch G, Zhang R, Siebert HC, Wilhelm J, Kaesser U, Dettmeyer RB, Klein H, Ishaque B, Rickert M, Schmitz G, Schmidt TA, Steinmeyer J (2015) Articular joint lubricants during osteoarthritis and rheumatoid arthritis display altered levels

and molecular species. PLoS ONE 10(5):e0125192. https://doi.org/10.1371/journal.pone.0125192
111. Kosinska MK, Mastbergen SC, Liebisch G, Wilhelm J, Dettmeyer RB, Ishaque B, Rickert M, Schmitz G, Lafeber FP, Steinmeyer J (2016) Osteoarthritis and cartilage 24(8):1470–1478, ISSN 1063–4584
112. Jónasdóttir HS, Brouwers H, Kwekkeboom JC, van der Linden HMJ, Huizinga T, Kloppenburg M, Toes REM, Giera M, Ioan-Facsinay A (2017) Targeted lipidomics reveals activation of resolution pathways in knee osteoarthritis in humans. Osteoarthritis Cartil 25(7):1150–1160. https://doi.org/10.1016/j.joca.2017.01.018
113. Van de Vyver A, Clockaerts S, van de Lest CHA, Wei W, Verhaar J, Van Osch GJVM, Bastiaansen-Jenniskens YM (2020) Synovial fluid fatty acid profiles differ between osteoarthritis and healthy patients. Cartilage 11(4):473–478. https://doi.org/10.1177/1947603518798891
114. Chen D, Yu H, Zhang Y, Huang Y (2022) Metabolomic analysis of extracellular vesicles from human synovial fluids. Microchem J 177:107257, ISSN 0026–265X. https://doi.org/10.1016/j.microc.2022.107257
115. Gobezie R, Kho A, Krastins B, Sarracino DA, Thornhill TS, Chase M, Millett PJ, Lee DM (2007) High abundance synovial fluid proteome: distinct profiles in health and osteoarthritis. Arthritis Res Ther 9(2):R36. https://doi.org/10.1186/ar2172
116. Steinbeck MJ, Nesti LJ, Sharkey PF, Parvizi J (2007) Myeloperoxidase and chlorinated peptides in osteoarthritis: potential biomarkers of the disease. J Orthop Res 25(9):1128–1135. https://doi.org/10.1002/jor.20400
117. Ritter SY, Subbaiah R, Bebek G, Crish J, Scanzello CR, Krastins B, Sarracino D, Lopez MF, Crow MK, Aigner T, Goldring MB, Goldring SR, Lee DM, Gobezie R, Aliprantis AO (2013) Proteomic analysis of synovial fluid from the osteoarthritic knee: comparison with transcriptome analyses of joint tissues. Arthritis Rheum 65(4):981–992. https://doi.org/10.1002/art.37823
118. Liao W, Li Z, Wang H, Wang J, Fu Y, Bai X (2013) Proteomic analysis of synovial fluid: insight into the pathogenesis of knee osteoarthritis. Int Orthop 37(6):1045–1053. https://doi.org/10.1007/s00264-012-1768-2
119. Åhrman E, Lorenzo P, Holmgren K, Grodzinsky AJ, Dahlberg LE, Saxne T, Heinegård D, Önnerfjord P (2014) Novel cartilage oligomeric matrix protein (COMP) neoepitopes identified in synovial fluids from patients with joint diseases using affinity chromatography and mass spectrometry. J Biol Chem 289(30):20908–20916. https://doi.org/10.1074/jbc.M114.554683
120. Liao W, Li Z, Li T, Zhang Q, Zhang H, Wang X (2018) Proteomic analysis of synovial fluid in osteoarthritis using SWATH-mass spectrometry. Mol Med Rep 17(2):2827–2836. https://doi.org/10.3892/mmr.2017.8250
121. Ahmed U, Anwar A, Savage RS, Thornalley PJ, Rabbani N (2016) Protein oxidation, nitration and glycation biomarkers for early-stage diagnosis of osteoarthritis of the knee and typing and progression of arthritic disease. Arthritis Res Ther 18(1):250. https://doi.org/10.1186/s13075-016-1154-3
122. Roller BL, Monibi F, Stoker AM, Bal BS, Cook JL (2016) Identification of novel synovial fluid biomarkers associated with meniscal pathology. J Knee Surg 29(1):47–62. https://doi.org/10.1055/s-0034-1394165
123. Corigliano A, Preianò M, Terracciano R, Savino R, De Gori M, Galasso O, Gasparini G (2017) C3f is a potential tool for the staging of osteoarthritis. J Biol Regul Homeost Agents 31(4 suppl 1):29–35
124. Shahid M, Manchi G, Brunnberg L, Raila J (2018) Use of proteomic analysis to determine the protein constituents of synovial fluid samples from the stifle joints of dogs with and without osteoarthritis secondary to cranial cruciate ligament rupture. Am J Vet Res 79(4):397–403. https://doi.org/10.2460/ajvr.79.4.397
125. Anderson JR, Smagul A, Simpson D, Clegg PD, Rubio-Martinez LM, Peffers MJ (2019) The synovial fluid proteome differentiates between septic and nonseptic articular pathologies. J Proteomics 202:103370. https://doi.org/10.1016/j.jprot.2019.04.020

126. Luo Q, Ji S, Li Z, Huang T, Fan S, Xi Q (2019) Effects of ultrasound therapy on the synovial fluid proteome in a rabbit surgery-induced model of knee osteoarthritis. Biomed Eng Online 18(1):18. https://doi.org/10.1186/s12938-019-0637-2
127. Kiapour AM, Sieker JT, Proffen BL, Lam TT, Fleming BC, Murray MM (2019) Synovial fluid proteome changes in ACL injury-induced posttraumatic osteoarthritis: Proteomics analysis of porcine knee synovial fluid. PLoS ONE 14(3):e0212662. https://doi.org/10.1371/journal.pone.0212662
128. Brophy RH, Cai L, Duan X, Zhang Q, Townsend RR, Nunley RM, Guilak F, Rai MF (2019) Proteomic analysis of synovial fluid identifies periostin as a biomarker for anterior cruciate ligament injury. Osteoarthritis Cartil 27(12):1778–1789. https://doi.org/10.1016/j.joca.2019.08.002
129. Timur UT, Jahr H, Anderson J, Green DC, Emans PJ, Smagul A, van Rhijn LW, Peffers MJ, Welting TJM (2021) Identification of tissue-dependent proteins in knee OA synovial fluid. Osteoarthritis Cartil 29(1):124–133. https://doi.org/10.1016/j.joca.2020.09.005
130. Jacobs CA, Keller LE, Zhang S, Fu Q, Hunt ER, Stone AV, Conley CEW, Lattermann C, Fortier LA (2023) Periostin regulation and cartilage degradation early after anterior cruciate ligament reconstruction. Inflamm Res 72(3):387–394. https://doi.org/10.1007/s00011-022-01678-9

Serhat Aladağ received his PhD degree in the Department of Bioengineering at Hacettepe University. He received BSc and MSc degrees in Biomedical Engineering from Erciyes University in 2013 and 2017, respectively. Serhat Aladağ, who works as Quality Manager at the National Control Laboratory of the Turkish Medicines and Medical Devices Agency, has studies in the fields of laboratory quality management systems, biopotential signal processing, clinical engineering, feature extraction and biomedical instrumentation. He continues Post-Doc research in the field of osteoarthritis specific biomarkers identification, bioaffinity chromatography, metabolomics, lipidomics and proteomics.

Ilayda Demirdiş graduated from Başkent University, Department of Molecular Biology and Genetics in June 2021. She started her master's degree in Molecular Biology Department of Hacettepe University Biology Department in February 2023 and is still continuing her education. Her research interests include molecular biology, metabolomics and nanobiotechnology.

Burcu Gökçal Kapucu is a research assistant at Hacettepe University, Department of Chemical Engineering. She completed her undergraduate education and master's degree in Chemical Engineering at Hacettepe University. She continues her doctorate in the same department. She is engaged in the synthesis of porous micro and nanoparticles, emphasizing structural customization for applications in materials science. Her research extends to biotechnological applications, exploring diverse uses of these particles in diagnostics and biomolecular interactions. Additionally, she focuses on surface modification techniques and the design of functional materials to enhance performance in various contexts.

Bilge Başak Fidan is a master student in the department of bioengineering at Hacettepe University. Her undergraduate is in Biology Education from Gazi University and Management Information Systems at Anadolu University. Studied for 1 year in the biotechnology department of the University of Cádiz, Spain during Erasmus. Her research focuses on LC-MS/Q-TOF analysis, muscoskeletal system diseases, proteomics, and metabolomics. She especially worked on biological samples including blood plasma, urine, cell culture, and tissue. She analyzed pharmaceutical samples including tablets, nano formulations, parenteral solutions, etc. micro-LC. She is the Assistant specialist in Project Management and Coordination office at TÜBİTAK.

Batuhan Erhan Aktaş , MD, is a third-year medical resident specializing in the Department of Sports Medicine at Hacettepe University Faculty of Medicine. He earned his medical degree from Gazi University Faculty of Medicine. Dr. Aktaş is actively engaged in research, with a focus on the biomechanics of the musculoskeletal system, exercise physiology, ultrasound imaging, and sports injuries. His dedication to advancing knowledge in these areas is evident through his participation in clinical and research studies.

Merve Gizer is a Ph.D. She is a researcher at the Middle East Technical University MEMS Center. She obtained her bachelor's degree from the Department of Biology and her M.Sc. from the Department of Bioengineering at Hacettepe University. During her M.Sc. degree, she focused on stem cell-assisted osteogenesis topics. She is currently working on male infertility using human induced pluripotent stem cells derived primordial germ cells and spermatogonial stem cells. She graduated from the Department of Stem Cell Sciences of Hacettepe University.

Ozan Kaplan is a Post-Doc researcher in the Faculty of Pharmacy, Department of Analytical Chemistry at Hacettepe University. His undergraduate from chemistry department, M. Sc. From analytical chemistry. His main research areas are mass spectrometry based metabolomics, lipidomics and proteomics where he employ method development and bioinformatical approaches to address various research questions.

Idil Yet received her PhD in Bioinformatics from King's College London/University of London in 2016. Before that, she was trained as a statistician and a computer scientist. She received a health informatics master degree from Middle East Technical University which she later on also worked as a postdoctoral fellow on detecting aneuploidies using single-cell sequencing. She is currently affiliated at Hacettepe University at the Department of Bioinformatics. Her research focuses on GWAS and analyzing several-omics and integrating them using AI/machine learning techniques.

Melis Şardan Ekiz has been a lecturer at Hacettepe University Advanced Technologies and Research Center since 2019. She graduated from the Department of Chemistry at Middle East Technical University in 2008 and gained her PhD degree in Material Science and Nanotechnology from Bilkent University in 2016. She is experienced in performing mass spectrometric analyses. Her qualifications include conducting mass-based omics studies and designing self-assembled peptide nanomaterials. She focuses on the development of smart materials that mimic the structure and function of natural biomacromolecules through the programmed self-assembly of synthetic peptides into supramolecular architectures.

Petek Korkusuz received her MD degree from Gazi University Faculty of Medicine; completed her residency in Histology and Embryology at Hacettepe University Faculty of Medicine. She received her Ph.D. from Gazi University Faculty of Medicine Department of Medical Pharmacology. She is a full time Professor in Hacettepe University Faculty of Medicine Department of Histology and Embryology. Her cross-appointments are in Nanomedicine and Nanotechnology, Bioengineering, Stem Cell Sciences and Oral Biology departments. Her research is focused on biomaterials and stem cells for tissue engineering and regenerative medicine, cannabinoids, cancer biology and male infertility based on somatic and germ stem cell niches.

Mustafa Çelebier is a Full Time Professor at Hacettepe University, Faculty of Pharmacy, Department of Analytical Chemistry. He graduated from Ankara University, Faculty of Science, Department of Chemistry. He has two master's degrees and one Ph.D. degree in Analytical Chemistry. His research focuses on pharmaceutical analysis using HPLC, CE and LC-MS. He performs LC-MS based metabolomics and proteomics applications on biological samples. He received the

Hacettepe University Young Scientist Award in 2017 and the Turkish Pharmacists Association Young Scientist Incentive Award in 2019.

Ömür Çelikbıçak , Ph.D., serves as a Professor in the Department of Chemistry at Hacettepe University. His academic journey is deeply rooted in Hacettepe University, where he attained his Bachelor of Science, Master of Science in Chemistry, and Doctorate in Physical Chemistry. His research expertise primarily revolves around the application of mass spectrometry in diverse fields such as Proteomics, Metabolomics, and Lipidomics. In addition to his profound interest in analytical techniques, Dr. Çelikbıçak has made significant contributions in the synthesis and characterization of synthetic polymers, particularly focusing on their development as smart materials. He is also renowned for his innovative methods in the purification and enrichment of biomolecules, utilizing the unique properties of these advanced materials.

Ali Tuncel is a full professor in the Department of Chemical Engineering at Hacettepe University. He completed his undergraduate education, master's degree and doctorate degree in Chemical Engineering at Hacettepe University. Receiving the TÜBİTAK encouragement award in 1998, Prof. Tuncel has extensively contributed to the fields of biotechnology, sensor technologies, and micro-nanotechnology. In his biotechnological applications, he employs advanced affinity techniques for isolation and detection of trace biomolecules, including the development of ELISA systems targeting tumor markers and blood protein determination. In the domain of nanocatalytic applications, Prof. Tuncel directs efforts toward energy production and water treatment, engaging in projects with international collaborations. His active synthesis of nanocatalysts for the degradation of organic pollutants and exploration of water oxidation catalysts for hydrogen generation highlights significant contributions to biomedical and environmental research.

Feza Korkusuz MD attended medical specification in Orthopedic Surgery and Traumatology at Gazi University Medical Faculty, Department of Orthopedic Surgery and Traumatology between 1987 and 1992 after graduating from Ankara University Faculty in Medicine in 1986. Between 1989 and 1990 he studied in Osaka University Faculty of Medicine, Department of Orthopedic Surgery from where he graduated with a postgraduate diploma. His clinical and research studies focused on Spine Surgery, Sports Medicine and Basic Research. In 1994 and in 2001 he became Associate Professor and Professor. In 1999 and 2000, he received the METU Prof. Dr. N. Parlar Foundation young investigator and the Scientific and Technological Research Council of Türkiye (TÜBİTAK) Promotion Awards. In 2008, he became an active member of the Association of Bone and Joint Surgeons (ABJS), and since 2006 he has been the corresponding and deputy editor of Clinical Orthopaedics and Related Research. In 2017, he became an active member to the Turkish Academy of Science (TUBA). He was the Advisor to the President of TÜBİTAK between 2018 and 2021. He currently is the Chairman of the Department of Hacettepe University Faculty of Medicine, Department of Sports Medicine.

The manufacturer's authorised representative in the EU is Springer Nature Customer Service Centre GmbH, Europaplatz 3, 69115 Heidelberg, Germany. If you have any concerns regarding our products, please contact ProductSafety@springernature.com

Printed and bound by CPI Group (UK) Ltd, Croydon, CR0 4YY
01/10/2026
02225882-0002